W0049983

Results and Problems in
Cell Differentiation

A Series of Topical Volumes in Developmental Biology

3

Editors

W. Beermann, Tübingen · J. Reinert, Berlin · H. Ursprung, Zürich

Nucleic Acid Hybridization in the Study of Cell Differentiation

Edited by H. Ursprung, Zürich

With contributions of

M. Birnstiel, Edinburgh · I. R. Brown, Calgary
R. B. Church, Calgary · R. C. C. Huang, Baltimore · K. H. Kim,
West Lafayette · I. Purdom, Glasgow · M. M. Smith, Baltimore
D. M. Steffensen, Urbana · H. Tobler, Zürich · R. Williamson,
Edinburgh · D. E. Wimber, Eugene

With 29 Figures

Springer-Verlag Berlin Heidelberg GmbH 1972

ISBN 978-3-662-22245-4   ISBN 978-3-540-37149-6 (eBook)
DOI 10.1007/978-3-540-37149-6

Originally published by Springer-Verlag Berlin Heidelberg New York in 1972
Softcover reprint of the hardcover 1st edition 1972
Library of Congress Catalog Card Number 70-188705.

Preface

The informational content of cells is encoded in the nucleotide sequences of their DNA. The restrictions on base pairing — A pairing with T (U), and G pairing with C — in nature assures the fidelity of replication of DNA in cell division, and of transcription. In the test tube, these restrictions can be exploited for ascertaining similarities and dissimilarities of nucleic acids of varying origin by measuring the kinetics of reassociation of polynucleotides to double-stranded molecules in DNA-DNA renaturation or RNA-DNA hybridization experiments, and by determining the thermal stability and other physical-chemical properties of the resulting hybrid molecules. This method has enormous potential for developmental biology. It offers a more direct approach to the ever-present question of the genetic identity of different cell types in an individual organism, and a more direct test of the hypothesis of differential gene function. It offers the possibility of localizing genes on chromosomes without the use of Mendelian genetics. It is an indispensable tool in the isolation, purification, and characterization of genes.

This volume brings together six articles by investigators actively working on various aspects of developmental biology who use nucleic acid hybridization as a tool in their research. Sound in theory, the method is in a honing phase as regards the technical detail. This is expressed in the hesitation with which some of the conclusions are rightly drawn.

We had originally intended to include a major chapter on the purely technical aspects of the method. The overriding difficulty facing our series of topical volumes — author synchronization — made this impossible. Nevertheless, we believe students of cell and developmental biology will find the book stimulating. For biochemists and molecular biologists it provides easy access to the literature of one of the most exciting approaches to the problem of cell differentiation.

Tübingen, Berlin, Zürich
April 1972

W. BEERMANN, J. REINERT,
H. URSPRUNG

Contents

Nucleic Acid Hybridization to Isolated Chromatin

by Ki-Han Kim

Hybridization of Nucleic Acids to Chromosomes

by Dale M. Steffensen and D. E. Wimber

Nucleic Acid Hybridization and the Nature of Chromosomal Protein Bound RNA

by RU CHIH C. HUANG and M. MITCHELL SMITH

Contributors

BIRNSTIEL, MAX, Dr., Institute of Animal Genetics, University of Edinburgh, Edinburgh (Scotland)

BROWN, I. R., Dr., Division of Medical Biochemistry, University of Calgary, Calgary, Alberta (Canada)

Church, R. B., Dr., Division of Medical Biochemistry, University of Calgary, Calgary, Alberta (Canada)

HUANG, RU CHIH C., Dr., Department of Biology, Johns Hopkins University, Baltimore, Maryland (USA)

KIM, KI-HAN, Dr., Department of Biochemistry, Purdue University, West Lafayette, Indiana (USA)

PURDOM, IAN, Dr., Royal Beatson Memorial Hospital, Glasgow (Scotland)

SMITH, M. M., Dr., Department of Biology, Johns Hopkins University, Baltimore, Maryland (USA)

STEFFENSEN, DALE M., Dr., Department of Botany, University of Illinois, Urbana, Illinois (USA)

TOBLER, HEINZ, Dr., Zoologisches Institut der Universität Zürich, Zürich (Switzerland)

WILLIAMSON, ROBERT, Dr., M. R. C. Epigenetics Research Group, Department of Genetics, University of Edinburgh, Edinburgh (Scotland)

WIMBER, D. E., Dr., Department of Biology, University of Oregon, Eugene, Oregon (USA)

The Problem of Genetic Identity of Different Cell Types

Heinz Tobler

Institute of Zoology and Comparative Anatomy, University of Zurich, Zurich

I. Introduction

An old dogma in biology states that all cells of a higher organism contain identical sets of genetic information in their nuclei. This classical concept is based on the fact that the cells of an organism are descendants of a single zygote and that the process of mitosis leads to equal assignment of the chromosomes and their genes to the daughter cells. Also direct measurements of the DNA content of various somatic cell types of an organism show equal amounts of DNA (VENDRELY, 1955). The only exceptions are provided by polyploid and polytene cells. It is generally believed, however, that these exceptional cases of increased DNA content per cell result from replications of the entire genome; such cells therefore would merely differ in a quantitative way from normal diploid somatic cells.

Aside from embryological evidence to be discussed later, the rule of DNA constancy is based on cytological and quantitative evidence only. This does not exclude the possibility that different cell types could differ from one another by differential gene amplification and/or alteration of some genes.

II. DNA-DNA Hybridization

Possible differences between the genomes of different cell types were investigated by McCARTHY and HOYER (1964) using DNA-DNA hybridization in competition experiments (Fig. 1). Labeled DNA from cultured mouse L-cells (a cell strain derived from connective tissue) was allowed to compete with unlabeled DNA's from various mouse tissues for complementary binding sites on DNA, which had been prepared from mouse embryos and entrapped in agar. All the unlabeled DNA's from different tissues were equally efficient competitors, indicating that there are no qualitative or quantitative differences in the DNA sequences between cells from different tissues. It was therefore concluded that cell differentiation is based on differential gene activity rather than on specific differences in the DNA content between different cell types. However, later findings cast doubt on the validity of this conclusion. It was shown that the sensitivity of the method used at the time is not very high; in fact, DNA's from different but closely related species could hardly be distinguished by this technique (HOYER et al., 1965). Also BRITTEN and KOHNE (1966; 1968) demonstrated very clearly that the genome of higher organisms is composed of repetitious as well as unique DNA sequences (Fig. 2). As an example, about 10% of the mouse genome consists of highly repetitious DNA base sequences present in about 1 000 000 copies in each diploid cell. A second fraction of the genome, amounting to about 15% of the total DNA, consists of families of related sequences on the order of 1000 to 100 000 copies. The bulk of the genome (70%) consists of unique DNA base sequences, which most likely represent genes occurring as single copies per haploid chromosome set.

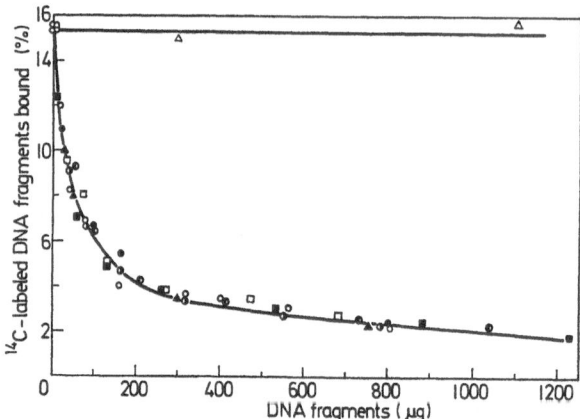

Fig. 1. Competition by unlabeled DNA fragments in the reaction of labeled DNA fragments with DNA agar. One microgram of ^{14}C-labeled DNA fragments (2500 cpm/µg) from mouse L cells was incubated with 0.5 gm of agar containing 60 µg of mouse embryo DNA in the presence of varying quantities of unlabeled DNA fragments from various mouse tissues, from mouse L cells, or from *B. subtilis*. The percentage of ^{14}C-labeled DNA fragments bound is plotted against the amount of unlabeled DNA present (From McCarthy and Hoyer, 1964) ○ Mouse L-cell; ● Embryo; ■ Kidney; □ Brain; ◑ Thymus; ◐ Spleen; ▲ Liver; △ *B. subtilis*

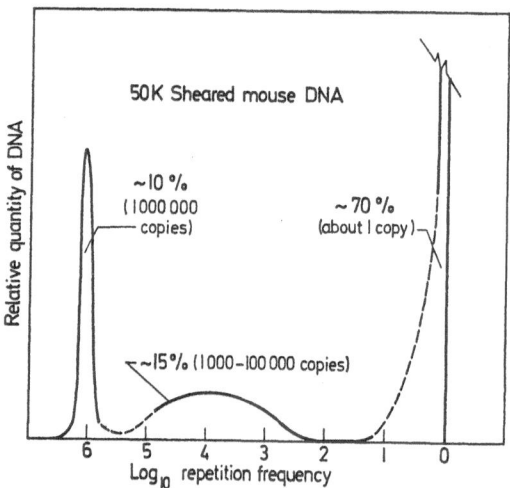

Fig. 2. Spectrogram of the frequency of repetition of nucleotide sequences in the DNA of the mouse. Relative quantity of DNA is plotted against the logarithm of the repetition frequency. These data are derived from measurements of the quantity and rate of reassociation of fractions separated on hydroxyapatite. The dashed segments of the curve represent regions of considerable uncertainty (From Britten and Kohne, 1968, Copyright 1968 by the American Association for the Advancement of Science)

Any given unique DNA sequence is enormously diluted in the large quantity of DNA in each cell of a higher organism, and therefore reanneals much more slowly than repetitious DNA sequences. From this fact, it became clear that the fraction of DNA which had been analysed in the DNA-agar hybridization experiments contained only the repetitious DNA sequences. Therefore, the statement that all cells of a higher organism contain qualitatively and quantitatively the same DNA sequences had to be modified. In the light of these new findings it could be concluded only that different cell types of an organism are qualitatively equal in their repetitious DNA. The possibility remained, however, that real qualitative differences exist among the unique DNA sequences from different cell types. Such differences would go undetected in the conventional hybridization technique.

III. Gene Amplification

A specific quantitative difference between the DNA content of different cell types is seen in the phenomenon of gene amplification in oocytes of amphibians and some other organisms (BROWN and DAWID, 1968; GALL, 1969). Hybridization of labeled ribosomal RNA (rRNA) to DNA extracted either from oocytes or from somatic tissues demonstrated that each oocyte of *Xenopus laevis* contains 1000—4000 copies of the chromosomal nucleolar organizer which is present as a single copy in each haploid chromosome set of a somatic cell. The nucleolar organizer itself comprises about 450 to 800 genes (rDNA) each for 28s and 18s rRNA (WALLACE and BIRNSTIEL, 1966; BROWN and WEBER, 1968). These extra copies of rDNA function in the germinal vesicle of the oocyte and were shown to be non-chromosomal. They are located in the more than 1000 nucleoli of the growing amphibian oocyte and are engaged in the synthesis of a great amount of rRNA (MILLER and BEATTY, 1969). This selective amplification of genes coding for 28s and 18s rRNA most likely serves the purpose of synthesizing large numbers of ribosomes to be used later during embryogenesis (BROWN and DAWID, 1968). Until now, selective amplification of rDNA has been found only in oocytes. DNA from sperm which carry out little rRNA synthesis, and from several somatic tissues differing greatly in the extent of their rRNA synthesis, showed the same rDNA content (RITOSSA et al., 1966; BROWN and DAWID, 1968). Also, large scale amplification of non-repeated DNA sequences which are expressed as RNA in different tissues of the calf can be ruled out, though it remains possible that low degree amplification of specific genes may occur in different cell types (KOHNE and BYERS, 1972).

The concept of selective gene amplification has recently been challenged by WALLACE et al. (1971), who suggest that the extra rDNA cistrons in the oocyte are derived from self-reproducing particles which are restricted to the germ-line. Whether the origin of these extra copies of rDNA in the nucleoli is completely independent of the chromosomal rDNA genes remains to be seen. The fact that in reciprocal crosses between *X. laevis* and *X. mulleri* the amplified rDNA is always of the *Xenopus laevis* type and therefore not maternally affected (BROWN and BLACKLER, 1972), and that extra copies of rDNA have never been found in the cytoplasm, are strong arguments in favor of the original gene amplification theory. Furthermore, chromosomal and amplified rDNA's are indistinguishable in their hybridization values with rRNA and in the stability of the hybrids formed (GALL, 1969; DAWID et al., 1970).

This is to be expected only if the two rDNA's are very similar or identical in their base sequence. If, however, the two rDNA's are not of a common origin, or are so only in an evolutionary sense, one might assume that mutations would have resulted in different rDNA types.

A phenomenon related to gene amplification in oocytes is found in the DNA-puffs of sciarid insects (PAVAN, 1965). Certain puffs of the polytene chromosomes of salivary glands show localized incorporation of ^{3}H-thymidine into the puffs, indicating DNA synthesis. In contrast to the gene amplification system in the oocytes, the extra DNA most likely remains in the chromosomes. The function of this selectively amplified DNA in the puffs is still unknown.

IV. Cell Specific Differences in Base Sequence

SCARANO et al. (1967) presented a highly speculative hypothesis according to which differentiation is based on specific alterations in the base sequence of the DNA. In their scheme the DNA of the germ-line cells remains unaltered, whereas differentiated cells result from mutation-like changes in base pairs brought about by the action of DNA-modifying enzymes. As differentiation proceeds, these enzymes would be sequentially synthesized. Such minor differences in the base sequence of DNA's from different tissues would probably remain undetected by the conventional hybridization techniques. One of these enzymes has indeed been found in sea urchin extracts. But proof has not been furnished thus far that it catalyzes the necessary nucleotide modifications in DNA. Also, the results from experiments in developmental biology argue strongly against the notion that cell differentiation is based on a series of directed mutations, as will be discussed below.

V. Embryological Evidence
A. Totipotency and Nuclear Transplantation

Undoubtedly the best evidence yet available for the qualitative identity of the genetic material among different cell types of an organism stems from embryological experiments, which showed that specialized cells remain totipotent with respect to their developmental capacities. All these examples indicate that differentiation does not depend on gene segregation, gene loss or irreversible gene repression and/or alteration. Single differentiated cells from a variety of normal and certain tumorous tissues of carrot, tobacco and other plant species are potentially able to give rise to an entire plant (STEWARD et al., 1964; BRAUN, 1968; TABEKE et al., 1971). Such remarkable regeneration capacities can even be achieved in a completely synthetic medium (VASIL and HILDEBRANDT, 1965; BACKS-HÜSEMANN and REINERT, 1970). In animals, it has never been demonstrated that single isolated cells from a differentiated somatic tissue retain their full developmental capacities. However, from GURDON's nuclear transplantation experiments it may be concluded that specialized cells must still contain the entire set of genes. Enucleated eggs of *Xenopus*, which have received a nucleus from differentiated intestinal cells of a feeding tadpole, are able to develop and give rise to normal and fertile frogs (GURDON and UEHLINGER, 1966). Recent experiments have shown that adult lung, skin, and kidney tissues contain nuclei, which upon injection into enucleated eggs can support the development of tadpoles (LASKEY and GURDON, 1970). It has always been noted in nuclear transplants in amphibians that only a very small number of transplanted nuclei are able to support normal development. Less than 2% of the intestinal epithelial cell nuclei promoted the formation of normal larvae, and only 0.3% of the originally transplanted nuclei gave rise to fertile adults. If serial nuclear transplants are added to the first transfers, however, then a significant proportion of 7% of intestine nuclei resulted in the formation of normal feeding tadpoles (GURDON, 1962).

From these experiments it seems unlikely that the "successful" nuclei stemmed from totipotent "reserve" cells that might be present in differentiated tissues. But the fact that the proportion of nuclei showing restricted developmental capacities always increases if the nuclei are derived from more advanced developmental stages, taken together with the above mentioned observation that only a few transplanted nuclei are able to support normal development, has been interpreted to mean that most nuclei do acquire irreversible restrictions during the process of cell differentiation (see DIBERARDINO and HOFFNER, 1970, for a recent discussion of this point of view of the nuclear transplantation studies). If this is the case, it appears more likely that they are based on irreversible gene repression rather than on gene segregation or alteration. However, the failures may also result from more trivial causes: 1) nuclei could become progressively more susceptible to experimental manipulations; 2) due to the small size of the donor cells, it is possible that an intact cell was transplanted rather than a liberated nucleus; 3) the donor nucleus may not be able to adjust its cell cycle to the rapid rate of cell divisions in the early cleavage stages.

B. Regeneration and Transdetermination

Further evidence for the notion that specialized cells normally contain the full set of genes comes from experiments which demonstrate that the determined or differentiated state of cells can be altered under certain experimental conditions. A well known example is Wolffian lens regeneration in some salamanders. After complete removal of the lens, differentiated cells from the dorsal iris lose their pigment granules, start to divide and subsequently redifferentiate into lens fiber cells, thereby synthesizing lens-specific proteins (YAMADA, 1967). Changes in cell types have also been demonstrated during normal growth and regeneration of *Hydra* (BURNETT, 1968), whereas it is still an unresolved problem whether or not true metaplasia occurs during limb regeneration of some vertebrates (see HAY, 1970). The discovery of transdetermination in *Drosophila* provides another example of a change in cell heredity (HADORN, 1965). Cells of imaginal disks kept in culture in abdomens of adult flies are able to switch from one determined state into another and to enter new developmental pathways.

None of the above mentioned examples of the reversibility of the differentiated state prove that the genetic material of the cells remained unchanged during the process of cellular differentiation; they merely show that it has not become irreversibly altered. There are, however, well known examples of irreversible and stable alterations of the genetic material in some cell types. First, certain terminally differentiated cells like mammalian erythrocytes, eye lens fiber cells or sieve elements in plants have lost their nuclear genetic material altogether. Second, chromosome diminution or elimination during early cleavage of some nematodes, insects, and crustaceans leads to cell populations which differ in their chromosomal content.

VI. Chromosome Diminution and Elimination

Probably the most famous case in which a portion of the genome is lost in some cell types but retained in others, is that of chromatin diminution and elimination of distal heterochromatin in *Ascaris* (Boveri, 1887). In early cleavage stages of the developing embryo, chromatin elimination takes place in the presumptive somatic cells but not in the germ-line, thus leading to cells with different amounts of DNA. In *Ascaris lumbricoides*, about 27% of the total DNA is eliminated from the somatic cells (Tobler et al., 1972). For *Parascaris equorum*, Moritz (1970) reported a value of more than 80% for eliminated DNA.

We have carried out molecular hybridization experiments in order to elucidate the composition of the DNA that is retained in the germ-line, but eliminated from

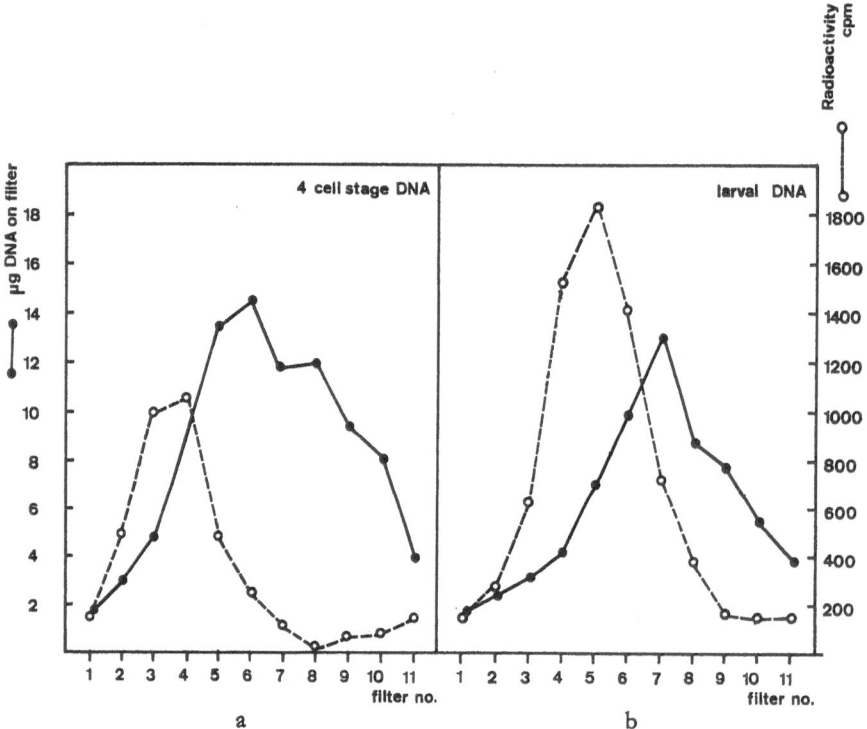

Fig. 3. Hybridization of ³H-labeled rRNA (158000 cpm/μg) from *Xenopus laevis* to CsCl-fractionated 4-cell stage (a) and larval (b) DNA from *Ascaris lumbricoides*. The total amount of DNA bound to all filters was 92 μg for 4-cell stages and 66 μg for larvae, respectively. All filters containing the DNA fractions were incubated together with interspersed blank filters in 7 ml of 2x SSC at 68° for 18 h in the presence of 5 μg of ³H-rRNA. ●——● μg DNA on filters, ○------○ cpm (From Tobler et al., 1972)

the somatic cells ("eliminated DNA") (Tobler et al., 1972). First we asked whether the eliminated DNA contains repetitious DNA sequences only, or whether unique sequences are also present in this portion of the genome. From the differences between the renaturation kinetics of spermatid DNA ("germline DNA") and somatic DNA from whole larval stages ("retained DNA") it became clear that the eliminated DNA is enriched for repetitious DNA sequences, compared with somatic DNA, but still contains a large fraction of unique DNA sequences.

If the oocyte of *Ascaris* contains selectively amplified rDNA genes, as is the case in amphibians (see p. 3), then the function of chromatin elimination might consist in the elimination of extra copies of rDNA from the somatic cells. Hybridization of rRNA with germ-line and somatic DNA clearly indicates that 4-cell stages are not enriched for rDNA genes compared with somatic cells (Fig. 3). The fact that germ-line DNA binds even less rRNA than does somatic DNA is consistent with the assumption that the eliminated DNA does not contain any ribosomal genes.

Obviously the most important question is whether the eliminated DNA differs in a qualitative or merely in a quantitative way from the retained DNA. A sequential hybridization experiment was performed in order to answer this question (Fig. 4). Because radioactively labeled DNA is not readily available from *Ascaris*, complementary RNA's (cRNA's) to both larval and spermatid DNA were transcribed *in vitro* using RNA polymerase from *E. coli*. These two cRNA's were hybridized

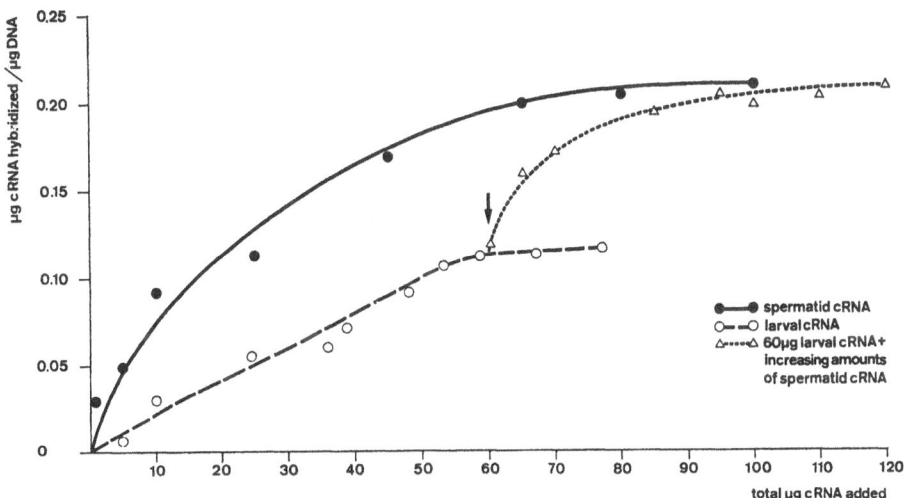

Fig. 4. Saturation of spermatid DNA from *Ascaris lumbricoides* with increasing amounts of complementary RNA (cRNA) derived from spermatid and larval DNA. Hybridization was carried out in 0.2 ml of 2 x SSC at 68° for 18 h using 1 µg DNA filters. The specific activities of the cRNA's were: spermatid cRNA = 6900 cpm/µg, larval cRNA = 2720 cpm/µg. After the saturation value for larval cRNA was determined, DNA filters were incubated as above with 60 µg larval cRNA and increasing amounts of spermatid cRNA. The results are indicated by the dotted line following the arrow (From TOBLER et al., 1972)

separately in increasing amounts to spermatid DNA. At saturation larval cRNA anneals with about 11% of the spermatid DNA, and spermatid cRNA with about 21%. If increasing amounts of spermatid cRNA are added after saturation of spermatid DNA with larval cRNA has been attained, the amount of hybridization rises again until the saturation level with spermatid cRNA is reached. This result clearly indicates that spermatid DNA does contain sequences which are no longer present in the larval DNA that was used as template to make cRNA. It may be concluded that the eliminated DNA contains repetitious DNA sequences which are qualitatively different from the retained repetitious DNA sequences of the somatic cells.

The hybridization experiments give no information as to the function of the eliminated DNA. However, some conclusions about this question have been reached from experiments with two species of *Cecidomyidae* (Diptera). Embryos of *Wachtliella* (Geyer-Duszyńska, 1966) and *Mayetiola* (Bantock, 1970) retain the full complement of some 40 chromosomes in the germ-line nuclei, but lose about 32 chromosomes in the presumptive somatic cells. Experiments were performed to induce chromosome elimination in the germ-line region, thus leading to prospective germ cells with only 8 chromosomes. Such individuals develop into sterile though otherwise normal adults. This result strongly suggests that the eliminated chromosomes contained at least some genetic information indispensible for normal gametogenesis.

VII. Conclusions

In this chapter we have presented evidence in favor of the basic idea that all cells of a higher organism contain identical sets of genetic information. We also pointed out that there are some important exceptions to this rule. Gene amplification in amphibian oocytes leads to cells with *quantitatively* different amounts of rDNA. Chromatin elimination in *Ascaris* results in *qualitative* differences in the fraction of repetitious DNA between germ-line and somatic cells. Only future research can elucidate whether the known examples really are the only ones in which different cell types of an organism differ quantitatively and/or qualitatively in their content of genetic information.

References

Backs-Hüsemann, D., Reinert, J.: Embryobildung durch isolierte Einzelzellen aus Gewebekulturen von *Daucus carota*. Protoplasma (Wien) **70**, 49—60 (1970).

Bantock, C. R.: Experiments on chromosome elimination in the gall midge, *Mayetiola destructor*. J. Embryol. exp. Morph. **24**, 257—286 (1970).

Boveri, T.: Über Differenzierung der Zellkerne während der Furchung des Eies von *Ascaris megalocephala*. Anat. Anz. **2**, 688—693 (1887).

Braun, A. C.: The multipotential cell and the tumor problem. In: Ursprung, H. (Ed.): The stability of the differentiated state, pp. 128—135. Berlin-Heidelberg-New York: Springer 1968.

Britten, R. J., Kohne, D. E.: Nucleotide sequence repetition in DNA. Carnegie Inst. Yearbook 1965 (Wash.) 78—106 (1966).

— — Repeated sequences in DNA. Science **161**, 529—540 (1968).

Brown, D. D., Blackler, A. W.: Gene amplification proceeds by a chromosome copy mechanism. J. molec. Biol. **63**, 75—83 (1972).

— Dawid, I. B.: Specific gene amplification in oocytes. Science **160**, 272—280 (1968).

— Weber, C. S.: Gene linkage by RNA-DNA hybridization. J. molec. Biol. **34**, 661—680 (1968).

Burnett, A. L.: The acquisition, maintenance, and lability of the differentiated state in *Hydra*. In: Ursprung, H. (Ed.): The stability of the differentiated state, pp. 109—127. Berlin-Heidelberg-New York: Springer 1968.

DAWID, I. B., BROWN, D. D., REEDER, R. H.: Composition and structure of chromosomal and amplified ribosomal DNA's of *Xenoxus laevis*. J. molec. Biol. **51**, 341—360 (1970).

DI BERARDINO, M. A., HOFFNER, N.: Origin of chromosomal abnormalities in nuclear transplants — a reevaluation of nuclear differentiation and nuclear equivalence in amphibians. Develop. Biol. **23**, 185—209 (1970).

GALL, J. G.: The genes for ribosomal RNA during oögenesis. Genetics **61**, 121—132 (1969).

GEYER-DUSZYŃSKA, I.: Genetic factors in oögenesis and spermatogenesis in *Cecidomyiidae*. In: DARLINGTON, C. D., LEWIS, K. R. (Ed.): Chromosomes Today, Vol. 1, pp. 174—178. Edinburgh: Oliver and Boyd 1966.

GURDON, J. B.: The developmental capacity of nuclei taken from intestinal epithelium cells of feeding tadpoles. J. Embryol. exp. Morph. **10**, 622—640 (1962).

— UEHLINGER, V.: "Fertile" intestine nuclei. Nature (Lond.) **210**, 1240—1241 (1966)

HADORN, E.: Problems of determination and transdetermination. In: Genetic control of differentiation. Brookhaven Symp. Biol. **18**, 148—161 (1965).

HAY, E. D.: Reversibility of the differentiated state. In: FRASER, F. C., McKUSICK, V. A. (Ed.): Congenital malformations. pp. 91—105. Amsterdam: Excerpta Medica 1970.

HOYER, B. H., BOLTON, E. T., McCARTHY, B. J., ROBERTS, R. B.: The evolution of polynucleotides. In: BRYSON, V., VOGEL, H. J. (Ed.): Evolving genes and proteins, pp. 581—590. New York-London: Academic Press 1965.

KOHNE, D. E., BYERS, M. J.: Studies on the amplification of expressed DNA sequences. (submitted for publication).

LASKEY, R. A., GURDON, J. B.: Genetic content of adult somatic cells tested by nuclear transplantation from cultured cells. Nature (Lond.) **228**, 1332—1334 (1970).

McCARTHY, B. J., HOYER, B. H.: Identity of DNA and diversity of messenger RNA molecules in normal mouse tissues. Proc. nat. Acad. Sci. (Wash.) **52**, 915—922 (1964).

MILLER, O. L., BEATTY, B. R.: Visualization of nucleolar genes. Science **164**, 955—957 (1969).

MORITZ, K. B.: DNS-Variation im keimbahnbegrenzten Chromatin und autoradiographische Befunde zu seiner Funktion bei *Parascaris equorum*. Verh. dtsch. Zool. Ges. **64**, 36—42 (1970).

PAVAN, C.: Nucleic acid metabolism in polytene chromosomes and the problem of differentiation. Brookhaven Symp. Biol. **18**, 222—241 (1965).

RITOSSA, F. M., ATWOOD, K. C., LINDSLEY, D. L., SPIEGELMAN, S.: On the chromosomal distribution of DNA complementary to ribosomal and soluble RNA. Natl. Cancer Inst. Monogr. **23**, 449—472 (1966).

SCARANO, E., IACCARINO, M., GRIPPO, P., PARISI, F.: The heterogeneity of thymine methyl group origin in DNA pyrimidine isostichs of developing sea urchin embryos. Proc. nat. Acad. Sci. (Wash.) **57**, 1394—1400 (1967).

STEWARD, F. C., MAPES, M. O., KENT, A. E., HOLSTEN, R. D.: Growth and development of cultured plant cells. Science **143**, 20—27 (1964).

TAKEBE, I., LABIB, G., MELCHERS, G.: Regeneration of whole plants from isolated mesophyll protoplasts of tobacco. Naturwissenschaften **58**, 318—320 (1971).

TOBLER, H., SMITH, K. D., URSPRUNG, H.: Molecular aspects of chromatin elimination in *Ascaris lumbricoides*. Develop. Biol. (in press), 1972.

VASIL, V., HILDEBRANDT, A. C.: Differentiation of tobacco plants from single, isolated cells in microcultures. Science **150**, 889—892 (1965).

VENDRELY, R.: The deoxyribonucleic acid content of the nucleus. In: CHARGAFF, E., DAVIDSON, J. N. (Ed.): The nucleic acids, Vol. 2, pp. 155—180. New York: Academic Press Inc. 1955.

WALLACE, H., BIRNSTIEL, M. L.: Ribosomal cistrons and the nucleolar organizer. Biochim. biophys. Acta (Amst.) **114**, 296—310 (1966).

— MORRAY, J., LANGRIDGE, W. H.: Alternative model for gene amplification. Nature New Biology **230**, 201—203 (1971).

YAMADA, T.: Cellular and subcellular events in Wolffian lens regeneration. In: MONROY, A., MOSCONA, A. A. (Ed.): Current topics in developmental biology, Vol. 2, pp. 247—283. New York-London: Academic Press 1967.

Tissue Specificity of Genetic Transcription

R. B. Church and I. R. Brown

Division of Medical Biochemistry and Biology, Faculty of Medicine,
University of Calgary, Calgary, Alberta

I. Introduction

The basic biochemical differences which exist between cells in animal tissues are reflections of characteristic patterns of proteins. The process(es) of differentiation, therefore, may be described as the process(es) whereby cells, presumably of identical genotype, develop into phenotypically distinct entities which reflect characteristic patterns of gene activity. The differential genetic expression which initiates these different protein patterns at different stages of development and in different tissues can involve regulation of the synthesis of the characteristic protein patterns at any or all of the following steps:

1) Differential RNA transcription including the possibility of specificity and efficiency differences in RNA polymerase;

2) Non-random stabilization and maturation of potential messenger RNA within the pool of unstable heterogeneous nuclear RNA including modifications in maturation, i.e., Poly-A termination;

3) Regulation of the transport of "mature" potential messenger RNA from the nucleus to the cytoplasm including physiological and hormone mediated non random alterations in this specific transport process;

4) Differential stabilities of messenger RNA molecules in the cytoplasm including modulation involving specific polypeptide processing;

5) Modulation of the translational efficiency of messenger RNA-ribosome complexes actively involved in protein synthesis.

6) Differential stability of the protein products.

There is evidence in the literature for the operation of all these processes in one cell type or another in animals. The differentiated pattern of proteins observed in any type of cell probably involves a dynamic equilibrium of all these processes, the differences in the stability of the differentiated state of any two tissues being governed by the susceptibility of each or all of the above phases of protein synthesis to external and cell-cell influences. Any given cell type, therefore, is the result of "differential gene activity" in response to specific external influences, such as inductive factors in embryological development or hormone induced changes in a differentiated target tissue. In this paper the analysis of differential gene activity is examined by the use of RNA/DNA reassociation techniques.

Differential gene activity has been fairly well documented in developing echino-derms and amphibians (see Davidson, 1968; Flickinger, 1970). Studies of the extent or complexity of RNA transcription from single copy sequences have been reported by Davidson and Hough (1969; 1971) and Firtel (1971). The complexity of ribosomal RNA sequences has recently been reviewed (Birnstiel et al., 1970). The amphibian oocyte, which contains all of the genetic information required for development, has a considerable amount of genetic information available for poten-tial protein synthesis stored in the different species of RNA molecules deposited during oogenesis (Denis, 1968). Compared to this wealth of information on sea urchins and amphibians, there is very little information available on the differential synthesis of RNA molecules during mammalian development.

II. Molecular Organization of the Mouse Genome

DNA/DNA and RNA/DNA reassociation is a powerful technique in the study of nucleic acid evolution (Britten and Kohne, 1968; Kohne, 1968); in the analysis of tissue distribution of RNA molecules (McCarthy and Hoyer, 1964; Church and McCarthy, 1967a, b; Smith et al., 1969, and Church, 1970); in studies of unstable nuclear RNA from mammalian cells (McCarthy et al., 1969). The technique of molecular hybridization with prokaryotic nucleic acids was first carried out in solu-tion by Nygaard and Hall (1964). In subsequent modifications of this technique, the single stranded DNA was trapped in agar by McCarthy and Hoyer (1964) and fixed on membrane filters by Gillespie and Spiegelman (1965) and Denhardt (1966) before hybridization, to simplify and speed up handling of the newly formed duplex. The technique of RNA/DNA hybridization using eukaryotic nucleic acids requires special consideration for proper interpretation of the results. The tremendous size of the genome in mammals presents problems in interpretation which are not usually of significance in viral and bacterial nucleic acid reassociation reactions (Church and McCarthy, 1968; McCarthy and Church, 1970).

The nuclear DNA of mammalian genomes contains an enormous amount of genome complexity. Genome complexity may be defined as the number of distinct DNA base sequences present in the haploid sperm complement of a species (Laird, 1971). The complexity of a genome is related to the genome size, approximately 6×10^{-12} g DNA per cell, in mammals. Britten and Kohne (1968) have established, on the basis of DNA/DNA reassociation kinetics of single stranded, sheared DNAs, that the genomes of higher organisms contain repeated (reiterated) and non repeated (unique or single copy) DNA base sequences.

Figure 1 shows the reassociation profile of ^{3}H labeled single copy DNA sequences in a mixture containing a vast excess of unlabeled total DNA. Previous reassociation studies showed that the mouse genome contains a very rapidly renaturing fraction equal to 6—9% of the genome. These sequences do not seem to be transcribed in any tissue or stage of development examined so far (Flamm et al., 1969); they are primarily located in the centromere and telomere regions of the chromosome (Pardue and Gall, 1969, 1970); They are also found associated with the heterochromatic regions of the chromosome (Yasmiueb and Yunis, 1969); and they include the AT satellite separated on CsCl gradients (Walker, 1969). Sequence analysis of this AT rich fraction suggests that the repeating units have a corrected complexity of 140 base pairs (Sutton and McCallum, 1971). This fraction renatures 10^6 times faster than would be expected if each sequence were present only once per mouse genome.

The second rapidly renaturing fraction, which makes up 20—25% of the genome, reassociates at an average rate from a few to 100000 times faster than expected for non repeated sequences. In light of the low RNA concentrations and the short

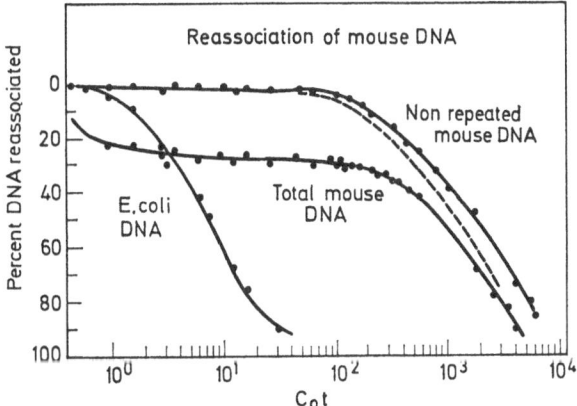

Fig. 1. Unlabeled DNA from mouse liver was mixed in 0.12 M PB with a low concentration of ³H-labeled nonrepeated mouse L-cell DNA. ¹⁴C-Labeled *E. coli* DNA was included as an internal renaturation standard. All DNA's were sheared to a single-strand molecular weight of about 120000. The mixture was heat-denatured (100° C, 10 min) and incubated at 60° C. Portions containing 100 µg of DNA were fractionated on hydroxyapatite columns to determine the percentage of denatured DNA remaining at various times. The C_0t values for mouse DNA's (total and unique) are based on the concentration of unlabeled mouse DNA. C_0t values for *E. coli* DNA were calculated from the concentration of this DNA. The dashed line represents the calculated second-order plot. The ratio of $C_0t_{1/2}$ values of mouse unique DNA to *E. coli* DNA is about 1000, an indication that the mouse unique DNA fraction represents sequences present, once per sperm (Laird, 1971). The mouse unique DNA does not reassociate with the fast-renaturing mouse sequences, indicating little repeated DNA contamination

reaction times it is probable that many of the hybridization estimates of transcriptional diversity only measure the RNA complementary to these repeated DNA base sequences (Church and McCarthy, 1968). There have been suggestions that the RNA transcripts complementary to repeated sequences are involved in gene regulation in animal cells (Britten and Davidson, 1969; Georgiev, 1969).

The remaining 60—70% of the mouse genome consists of sequences which renature at a rate constant which is consistent with the expected second order kinetics predicted for the mouse genome, assuming each sequence is present only once per haploid genome. The dashed line in Fig. 1 represents the theoretical second order plot for a haploid sperm genome size one thousand times larger than *E. coli*.

The reassociation plot of ³H-labeled fractionated single copy mouse DNA is also shown in Fig. 1. The single copy ³H-DNA was prepared from reassociation of sheared, (M.W. = 1,2 × 10⁵), denatured DNA (100° C for 10 min) in 0.12 M phosphate buffer (PB) at 60° C to a C_0t of 220. The incubation mixture was loaded onto a hydroxylapatite (HAP) column (0.12 M PB, 60° C) and the single stranded, single copies eluted with 0.12M PB. The HAP used was either prepared by the method of Bernardi (1969) or obtained from Cal Biochem and subsequently boiled at pH 11 before use. The HAP columns were never loaded beyond 70—100 μg DNA/ml, well below the capacity of this HAP to distinguish between single and double stranded nucleic acid duplexes.

The repeated reassociated double stranded duplexes can be washed from the HAP column with 0.5M PB or melted off by increasing the column temperature. The ^{3}H single stranded, single copy DNA was concentrated by ultrafiltration and dialysis, mixed with an excess of unlabeled total DNA, denatured and renatured in 0.12M PB, 60° C, to C_0t 10^4. In control experiments the single stranded ^{3}H DNA sequences eluted at 0.12M PB were concentrated, heated to 100° C for 10 min., renatured to C_0t 220 and eluted from hydroxylapatite; more than 98% of the DNA was again eluted in the 0.12M PB wash. It should be emphasized that the molecular weight of the DNA must be reduced to 1.2×10^5 so that the physical separation of repeated from single copy sequences can be achieved with HAP chromatography. As illustrated in Fig. 1 the single copy ^{3}H DNA reassociated slightly more slowly than a single second order reaction component would with a reaction rate about 1000 times slower than that of *E. coli*. The total unlabeled mouse DNA renatured at rates consistent for a genome containing more than one component. These experiments suggest that the purified single copy DNA sequences of this molecular weight are essentially free of repetitive sequences. This fraction does, however, contain at least one copy of each sequence of the repeated DNA because of the random shear sites (Britten, 1969). It must also be emphasized that the distinction between repeated and non-repeated DNA is somewhat arbitrary depending upon the criteria used in the reassociation reaction and the molecular weight of the DNA (Church and McCarthy, 1968).

Criteria are defined here as a measure of the stringency of the conditions of the reassociation reaction (McCarthy and Church, 1970). By varying the conditions of reassociation, or criteria, it is possible to delineate the degree of base pair mismatching in a particular experiment. Laird et al. (1969), on the basis of the analysis of melting profiles of duplexes, established that 1.5% base pair mismatching was reflected in a 1° C drop in the T_m of the melting profile of the reassociation duplex. Theoretically, with the ultimate criteria all sequences could conceivably behave as single copy sequences. In the experiments described here about 65% of the DNA sequences in the mouse genome appear in the single copy 0.12M PB fraction.

III. Transcriptional Activity of the Repeated DNA Sequences

In the design of RNA/DNA reassociation reactions for mammalian nucleic acids the criteria of reassociation are very important. The presence of DNA sequences represented by families of similar but not identical DNA sequences of varying repetition frequency means that RNA/DNA hybridization rarely if ever displays cistron specificity (Church and McCarthy, 1968). Therefore, two single stranded DNA fragments which were complementary strands of the native duplex will seldom reassociate with each other since the probability of their reassociating with other members of the DNA base sequence family is greater (McCarthy and McConaughy, 1968). Similarly it is unlikely that RNA transcripts from DNA sequences which are members of these families will ever form reassociation duplexes from the original transcription site (CHURCH and McCARTHY, 1968). The reassociation criteria, salt concentration, concentration of organic solvents, temperature, nucleic acid fragment size and reaction time will determine the degree of DNA/RNA base sequence mismatching which will be permitted in duplex formation. The degree of sequence relatedness is measured by the extent of base mismatching which can be determined by the T_m of the duplexes (LAIRD et al., 1969). The choice of particular reassociation criteria for an experimental system utilizing mammalian repeated DNA base sequences is an arbitrary one which allows a minimum percentage of base pair mismatching, measures a portion of the genome and is consistent with the specific activity of the labeled nucleic acid (McCARTHY and CHURCH, 1970). The higher the temperature and lower the salt concentration, the more specific is the base pairing required to form a stable duplex. This implies that both extent and rate of reaction may be greatly reduced and may be incompatible with the specific activity of the labeled component particularly for *in vivo* animal isotope labeling. Organic solvents which act as denaturants allow lower temperatures of incubation (20 to 50° C), avoid the extensive thermodegradation which can occur in reactions above 50° C and allow special reassociation reactions with aminoacyl-tRNA hybridization (NASS and BUCK, 1970). The retention of single stranded DNA, especially sheared DNA fragments, by membrane filters is greatly enhanced by low reaction temperatures. BONNER et al. (1967) reported the use of formamide in reassociation studies and more recently McCONAUGHY et al. (1969) described the relationship of formamide concentration to the thermal stability of DNA/DNA reassociation duplexes. The T_m of double stranded DNA duplexes was reduced 0.72° C for every additional 1% formamide added to the nucleic acid reassociation mixture in the temperature range from 45 to 90° C. Studies of mammalian nucleic acid reassociation utilizing formamide are based on the assumption that the effect of the formamide on double stranded nucleic acid structure disruption is the same as its effect on duplex formation from single stranded nucleic acids. It should be noted that this assumption has not been tested directly.

Transcription of messenger RNA involves some as yet poorly defined processes whereby some sequences of the large, rapidly labeled, unstable nuclear RNA are selectively cleaved and transported in a nonrandom manner from the nucleus to the cytoplasm, presumably to act as templates in the translational process. Cytoplasmic transfer of unstable nuclear RNA is quantitatively inefficient and decidedly nonrandom (McCARTHY et al., 1969). In addition, in any given tissue only a fraction of the information encoded in the genome is transcribed. Current theories suggest that histones act as repressors of transcription of genetic loci but the specificity for recognition of a particular genetic loci is a property of acidic proteins (PAUL, 1970; DUERKSEN and McCARTHY, 1971). Isolated chromatin can be shown to exhibit a very restricted transcriptional activity *in vitro*. In experiments reported by SMITH et al. (1969) increasing amounts of labeled RNA synthesized *in vitro* using either liver chromatin or purified liver DNA were hybridized with mouse DNA. Two and one half times the number of DNA sequences were found to be complementary to RNA synthesized from a purified DNA template as compared to those RNA molecules transcribed from liver chromatin by homologous polymerase. The data suggest that 0.14 to 0.16% of the DNA in the liver chromatin is available as a template for homologous RNA polymerase.

The restrictions imposed on transcription in each tissue are detectable by *in vitro* chromatin experiments of the type shown in Fig. 2. Labeled RNA synthesized *in vitro* from mouse liver and kidney chromatin utilizing homologous mouse RNA polymerase was compared to the spectrum of RNA molecules in the same tissue *in vivo*. The homologous liver RNA population proves to be the most efficient com-

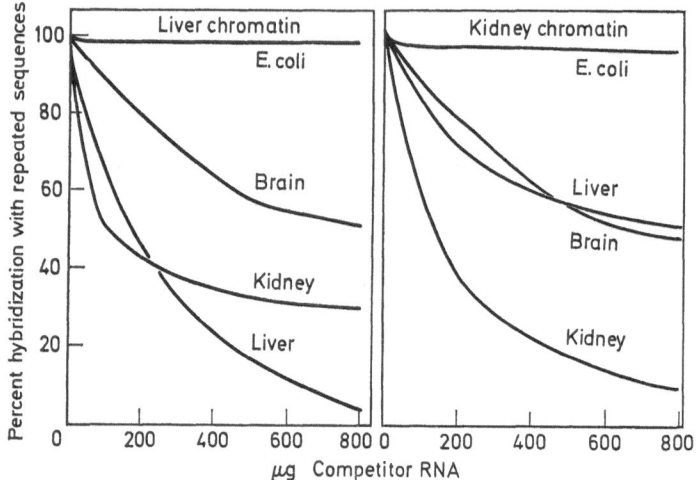

Fig. 2. Tissue specificity of RNA synthesized from chromatin originating from liver or kidney using mouse RNA polymerase. (1) RNA was synthesized *in vitro* from a mouse liver chromatin template using mouse RNA polymerase. The reaction was run in 5 ml buffer at 37° for 10 min and the RNA was extracted with hot phenol. RNA isolated from various tissues was used as competitor for the hybridization of the *in vitro* synthesized RNA to DNA. Hybridization with 12 μg of DNA in 0.2 ml of 2X SSC at 67° for 18 h; 384 cpm was hybridized in the absence of competitor. (b) RNA was synthesized *in vitro* from a mouse kidney chromatin template using mouse RNA polymerase. The reaction was run in 5 ml buffer at 37° for 10 min and the RNA was extracted with hot phenol. RNA isolated from various tissues was used as competitor for the hybridization of the *in vitro* synthesized RNA to DNA. Hybridization was with 12 μg of DNA in 0.2 ml of 2X SSC at 67° for 18 h; 336 cpm was hybridized in the absence of competitor (From Smith et al., 1969)

petitor while the kidney and brain RNA preparations exhibit partial homologies. Similar results can be seen for the RNA/DNA reassociation system utilizing RNA synthesized from kidney chromatin *in vitro*. Each chromatin template allows transcription of a specific array of RNA molecules characteristic of a tissue type. Smith et al. (1969) and Paul (1970) have shown that the synthesis of new RNA transcripts from repeated sequences is initiated *in vitro* and that all RNA molecules complementary to repeated sequences in the genome are present *in vitro*.

Experiments designed to analyze the spectrum of RNA transcripts complementary to repeated sequences in foetal, adult and regenerating mouse liver have been carried out (Church and McCarthy, 1967a and b). RNA/repeated DNA sequence-reassociation studies have also been used to examine the spectrum of RNA molecules present in estrogen stimulated uterine endometrium (Church and McCarthy, 1970), hormone stimulated avidin synthesis in chicken oviduct (Hahn and Church, 1969) and in many other systems. It is evident from the studies that populations of RNA molecules complementary to the repeated DNA sequences of the genome reflect differential gene activity in a very tissue specific and stage specific manner. The significance of the specific transcriptional patterns of repeated DNA sequences may be related to some regulatory function (BRITTEN and DAVIDSON, 1969, GEORGIEV, 1969), to the amplification of special structural genes such as those for histones (KEDES and BIRNSTIEL, 1971) or to synthesis of unstable nuclear RNA with some precursor relationship to cytoplasmic messenger RNA (McCARTHY et al., 1969).

RNA/DNA reassociation studies have also been a useful tool in the analysis of RNA information transfer from the nucleus to the cytoplasm in mammalian cells. This unstable nuclear RNA is rapidly synthesized, has a short half life (SHERRER and SPOHR, 1970), and therefore is not all a precursor of cytoplasmic messenger RNA. PENMAN et al. (1970) have suggested that there is no simple precursor relationship between unstable nuclear RNA and the cytoplasmic messenger RNA in mammalian cells. In contrast, the RNA/DNA reassociation studies carried out by McCARTHY et al. (1969), utilizing mouse L-cells, mouse liver and rabbit endometrium suggest that approximately 10% of the unstable nuclear RNA sequences are transported to the cytoplasm presumably to act as presumptive messages (SHEARER and McCARTHY, 1967).

The RNA molecules isolated from nuclei of rabbit uterine endometrium cells or mouse L-cells do not seem to sediment with any particular velocity in sucrose gradients since they are detected across the whole gradient. Studies of RNA transcriptional activity in mammalian cells employing the filter technique, or *low* C_0t RNA/ DNA hybridization, have provided useful information for qualitative changes in RNA synthesis from repeated sequences in many systems which have recently been reviewed (WALKER, 1969; McCARTHY and CHURCH, 1970; PAUL, 1970; KENNELI, 1971). McCARTHY and coworkers have presented data concerning the use and interpretation of these methods (1968a, b). The thermal stability of RNA/repeated DNA duplexes indicate incomplete base pairing which is in contrast to that expected for RNA/unique DNA duplexes. Therefore, while important conclusions can be drawn from high criteria RNA/repeated DNA reassociation experiments, quantitative information regarding the complexity of RNA transcription during tissue differentiation is most readily assayed by reassociation with DNA fractionated to remove the repetitious sequences.

IV. Transcriptional Activity of Single Copy DNA Sequences
A. Methods

The rate of hybridization depends on the effective concentration of complementary DNA sequences. Therefore, at low RNA concentrations and short reaction times (16—30 h), independent of DNA fragment size, buffer, Na^+ concentration, or reaction temperature, only those sequences present in multiple copies will have a chance to react. To analyze transcriptional activity of the non repeated, single copy sequences of the complex animal genome very high concentrations of nucleic acids or very long reaction times (or both) are required to allow RNA transcripts complementary to single copy DNA sequences the chance to react (McCARTHY, 1967; BRITTEN, 1969; LAIRD, 1971). RNA/DNA reassociation products of such *high* C_0t reactions have been analyzed in several laboratories (Gelderman et al., 1971; Davidson and Hough, 1969; 1971; Laird, 1971; Firtel, 1971; Brown and Church, 1971; Melli et al., 1971).

The RNA/DNA hybrid product formed under *high* C_0t reaction conditions has a T_m which approaches DNA/DNA reassociation duplexes (Brown and Church, 1971) indicating a minimum amount of mismatching compared to high criteria RNA/repeated DNA duplexes (Church and McCarthy, 1968). To obtain the high concentrations of nucleic acids required for high C_0t reassociation analysis of mammalian nucleic acids, RNA/DNA reassociation reactions in solution are required, since the insolution reaction rate is 20 times that of the filter reaction rate for the same system (McCarthy, 1967).

There are two methods presently in use for preparation of high C_0t duplexes. These involve either a large excess of unlabeled RNA with a small amount of highly labeled DNA (RNA-driven reaction) first used by Kohne (1968), or a very large excess of DNA in reaction with a small amount of highly labeled RNA ("DNA driven reaction") first used by Britten (1970) and extended by Bishop (1971). The RNA-driven reaction has been used to analyze the complexity of RNA transcription in the whole mouse embryo (Gelderman et al., 1971), slime mold development (Firtel, 1971) and mouse tissues including brain (Brown and Church, 1971; Hahn and Laird, 1971; Smith, 1968). The DNA-driven procedure has been used for analysis of duck erythrocyte mRNA (Bishop, 1971) and for ribosomal cistrons (Birnstiel et al., 1970).

The method of choice in the investigation of a biological system depends on whether one can label DNA or RNA to very high specific activities ($> 5 \times 10^5$ dpm/µg). *In vivo* labeled RNA usually calls for the DNA-driven reaction. In our laboratory the RNA is sheared to about 6 S or less and denatured at 100° C for 10 min in capillary tubes with the DNA before reassociation. This allows high concentrations (20 mg/ml) to go into solution and minimizes RNA secondary structure.

The problems of the RNA-driven reaction lie in the difficulty of HAP in separating single stranded from double stranded molecules and the various intermediate forms. This is best overcome in our laboratory by the pretreatment of the RNA/DNA hybrid with the "single stranded nuclease" of Fraser et al. (1970). No free labeled DNA sequence "ends" should be left on the double stranded complex for the radioactivity assay of the duplex. The extended incubation times imply that all glassware and nucleic acid preparations must be prepared with extreme care. Cations, which activate nuclease activity can be removed by Chelex treatment (Davidson and Hough, 1969). SSC, formamide, protein and HAP fine particles affect the characteristics of HAP elution and its capacity to distinguish between single and double stranded molecules. It has been our experience that formamide and high salt yield much slower rates of reaction than do comparable aqueous reactions. Therefore, all single copy DNA sequence hybrids are run in aqueous solutions at high temperature in our laboratory. It is absolutely essential to ensure that the labeled DNA contain no RNA but more specifically that the unlabeled RNA be free of contaminating DNA. Our standard control reaction involves the addition of alkali or ribonuclease to the reaction mixture. The elimination of the RNA stops the reaction, as shown in Fig. 3.

The DNA-driven system, provided one can obtain very high specific activity RNA, makes it possible to measure reiteration frequency of DNA sequences complementary to purified messenger RNA (MELLI et al., 1971). A high degree of purity of nucleic acids is not required since the proportion of RNA hybridized is assayed. The excess of DNA over RNA has to be 10^8 or more for the complex DNA of the mammalian genome in order for the reaction to go to completion. The problems of the method involve the isolation of sufficiently highly labeled RNA, obtaining and keeping the very high DNA concentration in solution, both of which will reduce the extent of reaction compared to the calculated theoretical maximum.

B. Transcription in Adult Tissues

The kinetics of reassociation of mouse liver, kidney, spleen and brain RNA in an RNA-driven reaction are presented in Fig. 3. The reaction cannot be attributed to traces of DNA contaminating the RNA preparation since the reaction is completely eliminated by alkali or RNAse treatment prior to incubation. The degradation of RNA under the reaction conditions used is $< 5\%$ based upon the exclusion of RNA in the incubation mixture from Sephadex G200 after reaction times of up to six weeks. The RNA used in these experiments was unlabeled hence no direct estimate can be made of the total content of RNA in the DNA/RNA reassociation duplex. The melting profiles of the DNA/RNA reassociation hybrids shown in Fig. 3 had $T_{m's}$ of more than 80.5° C, in close agreement with that found by HAHN and LAIRD (1971).

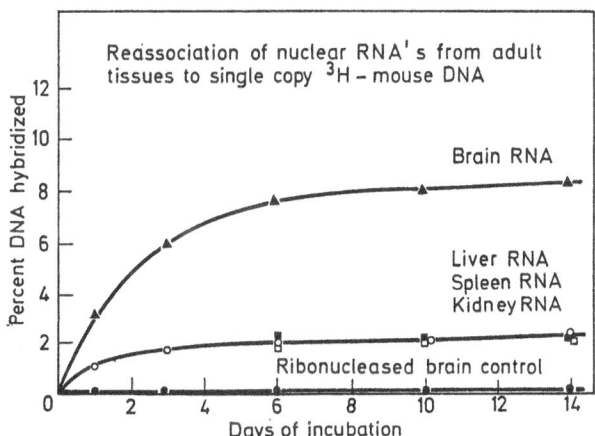

Fig. 3. The kinetics of reassociation of nuclear RNA's isolated from adult mouse liver, kidney, spleen and brain in reaction with ³H labeled mouse single copy DNA. The reassociation reaction mixture including 0.5 µg of single copy ³H DNA ($C_0 t$ 220); 1300 µg of brain (▲); 300 µg liver (■); 300 µg kidney (○) or 300 µg spleen (□) RNA in 50 µl of 0.12M PB in sealed capillary tubes were heated to 100° C for 10 min and incubated at 60° C for the times indicated. The reaction was stopped and the mixture treated with FRASER et al. (1970) nuclease or RNase. The reaction mixture was usually passed over a Sephadex G-200 column before the exclusion peak was loaded onto a hydroxyapatite column in 0.03 or 0.12M PB at 60° C. Any single stranded nucleic acid remaining was eluted with 0.12M PB at 60° C. The double stranded duplexes were either melted from the column by sequential temperature steps or eluted with 0.5M PB. The percentage of single copy ³H-DNA hybridized to RNA at a given time was calculated after subtraction of the 0.5% background obtained in control experiments with prior treatment with ribonuclease or alkali to remove the RNA (BROWN and CHURCH, 1971)

Approximately 8% of th ³H mouse DNA is complementary to total brain RNA. The complexity of the spectrum of RNA molecules from other tissues is about 2%. The apparent saturation values found in these experiments probably reflect minimal complexities of RNA transcription in these tissues. The values reflect the diversity of cell types found in the various tissues examined. We have not yet examined the complexity of RNA transcription from a single cell type. The fact that several body tissues show similar saturation values does not necessarily imply that the spectrum of RNA molecules in these tissues is identical. Preliminary experiments indicate that there is no more than a 50% overlap in total RNA complexity between these various tissues. Additivity experiments utilizing liver and spleen RNA resulted in apparent saturation values which exceeded those obtained for the homologous RNA preparations alone under similar reassociation conditions. A large amount of transcriptional overlap in different tissues of the body would be expected since many enzymes and structural proteins must be common to many cell types in different tissues.

C. Transcription in Embryos

We have recently become interested in studying the transcriptional patterns in the mouse embryo, utilizing the *in vitro* culture technique developed by Brinster (1965) with 50% serum for preimplantation stages of development. The results of these experiments (Church, 1970) and those of Woodland and Graham (1969) indicate that RNA synthesis has commenced by the two cell stage of the mouse embryo *in vitro*. Fig. 4 presents a summary of our analysis of transcriptional complexity for brain and liver tissue and the total mouse embryo during development. The extent of transcription of RNA molecules complementary to the repeated DNA sequences obtained under the high criteria used is, at best, a crude estimate, as indicated previously (Church and McCarthy, 1968). The extent of RNA transcription complementary to fractionated single copy DNA sequences is a minimal estimate of total

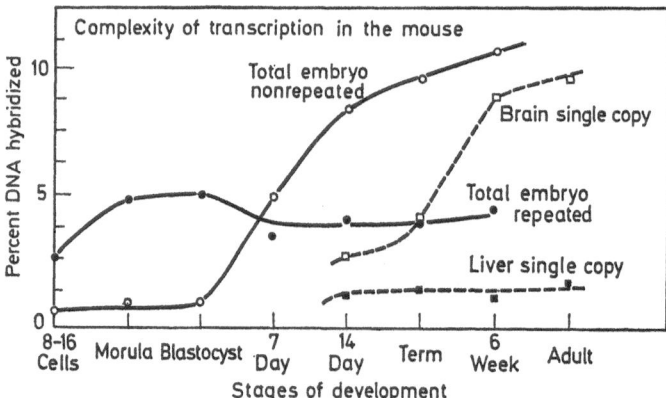

Fig. 4. The complexity of transcription of total nuclear RNA isolated from whole mouse embryos, liver and brain. The reaction of repeated DNA sequences with RNA isolated from all embryonic tissue was carried out by the filter technique as described previously (Church, 1970). The reaction of non repeated ³H DNA ($C_0 t$ 220) with unlabeled nuclear RNA isolated from embryos (○, ●), liver (■) and brain (□) was carried out as outlined in Fig. 3 (Brown and Church, 1971). The time scale is not linear and is diagrammatic

genetic activity. The conditions of reassociation are presented elsewhere (CHURCH, 1970; BROWN and CHURCH, 1971).

The extent of reaction of total mouse embryo RNA with the repeated sequences of the genome increases rapidly during the preimplantation stages of embryonic development before plateauing for the rest of embryonic development. It must be stressed that the true meaning of the extent of reaction of RNA transcripts complementary to repetitive DNA sequences is not clear. It is not possible to separate the contribution of RNA transcription from a few large DNA sequence families from the total transcriptional activity of a large number of smaller DNA base sequence families. Therefore, conclusions regarding the extent of transcription from repeated DNA sequences cannot be drawn.

Total embryonic RNA, as measured by its reassociation properties with the fractionated single copy DNA, shows increasing complexity of RNA transcription with embryonic development. Approximately 1% of the single copy portion of the genome is transcribed in the preimplantation stages of mouse development followed by an increase to approximately 10% by parturition. The percentage of the genome transcribed in brain tissue increases dramatically through embryonic development until maturity of the animal. The proportion of the total animal's gene activity accounted for by neural tissue increases steadily until, by six weeks of age, the complexity of transcription in the neural tissue of the mouse is equal to that found for the total mouse embryo at parturition. This estimate of genetic complexity of RNA transcription in the mouse brain is equivalent to more than 300000 different DNA sequences, each 1000 nucleotides in length (HAHN and LAIRD, 1971). The transcriptional complexity in the entire mouse must reflect the differential gene activity of the numerous cell types found in the mouse brain. In contrast to the genetic activity in neural tissue, parallel studies with liver RNA isolated from embryonic and postpartum mice of different ages did not show significant changes in the extent of transcriptional complexity.

V. Conclusions

Our experiments suggest that there is no preferential transcription of either repeated or non repeated base sequences in the mouse genome. However, differences seem to occur during differentiation. The estimates of transcriptional complexity presented in these studies in mouse tissues represent minimal values of genetic activity. The potential genetic information available in the mouse genome is far greater. Much of the complexity of transcription must be represented in the unstable heterogenous RNA populations found in the cell nucleus. The present experiments assay the accumulated complexity in RNA molecules present at the time of isolation. The precise meaning of the surprising complexity of transcription in neural tissue in comparison to the other body tissues presented in this study cannot be meaningfully interpreted at this time. Further experiments utilizing homogeneous cell populations are required to determine the contribution that glial, neuronal and other cell types make to the extent of genetic activity. It would appear that in the evolution of animals efficiency of RNA transcription has not been of utmost importance since so small a portion of the total transcriptional activity can presently be equated to messenger RNA activity in the cytoplasm of animal cells. Evolution may select for transcription which has maximal flexibility, thereby providing the cell with a ready mechanism to respond to different stimuli and tissue environments. The complexity of RNA transcription in the nucleus of cells in different tissues may provide a pool of potential messenger RNA's from which the cell can select specific members for transport to the cytoplasm for use as templates in characteristic protein biosynthetic patterns.

Acknowledgement

We wish to thank the National Research Council of Canada for operating grant support and the Medical Research Council for a Post Doctoral Fellowship (I.B.).

References

Bernardi, G.: Chromatography of nucleic acids on hydroxyapatite. I. Chromatography of native DNA. Biochim. biophys. Acta (Amst.) 174, 423—434 (1969).

Birnstiel, M. L., Chipchase, M., Speirs, J.: The ribosomal RNA cistrons. In: Davidson, J. N., Cohn, W. E. (Eds.): Proc. Nucl. Acid Res. Mol. Biol. Vol. 11, pp. 351—390 (1970).

Bishop, J.: Personal communication (1971).

Bonner, J., Kung, G., Bekhor, I.: A method for the hybridization of nucleic acid molecules at low temperature. Biochemistry 6, 3650—3656 (1967).

Brinster, R.: Studies on the development of mouse embryos *in vitro*. J. Reprod. Fertil. 10, 227—234 (1965).

Britten, R. J.: Repeated DNA and transcription. In: Hanly, E. W. (Ed.): Problems in Biology: RNA in Development, pp. 187—210. Univ. of Utah Press.

— Davidson, E. H.: Gene regulation in higher cells. Science 165, 349—357 (1969).

— Kohne, D. E.: Repeated sequences in DNA. Science 161, 529—540 (1968).

Brown, I. R., Church, R. B.: RNA transcription from nonrepetitive DNA in the mouse. Biochim. Biophys. Res. Commun. 42, 850—856 (1971).

Church, R. B.: In: Congenital Malformations. Fraser, F. C., McKusick, V. A. (Eds.): Differential gene activity. pp. 19—28. New York: Excerpta Medica 1970.

— McCarthy, B. J.: Ribonucleic acid synthesis in regenerating and embryonic liver. I. The synthesis of new species of RNA during regeneration of mouse liver after partial hepatectomy. J. molec. Biol. 23, 459—475 (1967a).

— — Ribonucleic acid synthesis in regenerating and embryonic liver. II. The synthesis of RNA during embryonic liver development and its relationship to regenerating liver. J. molec. Biol. 23, 477—486 (1967b).

— — Related base sequences in the DNA of simle and complex organisms. II. The interpretation of DNA/RNA hybridization studies with mammalian nucleic acids. Biochem. Genetics 2, 55—81 (1968).

— — Unstable nuclear RNA synthesis following estrogen stimulation. Biochim. biophys. Acta (Amst.) 199, 103—114 (1970).

Davidson, E. H.: Gene activity in early development, pp. 247—346. New York: Academic Press 1968.

— Hough, B.: High sequence diversity in the RNA synthesized at the lampbrush stage of oogenesis. Proc. nat. Acad. Sci. (Wash.) 63, 342—349 (1969).

— — Genetic information in oocyte RNA. J. molec. Biol. 56, 491—506 (1971).

Denhardt, D. T.: A membrane-filter technique for the detection of complementary DNA. Biochim. biophys. Res. Commun. 23, 641—645 (1966).

Denis, H.: Role of messenger ribonucleic acid in embryonic development. In: Advances in Morphogenesis, Vol. 7, pp. 115—150. New York: Academic Press 1968.

Duerksen, J. D., McCarthy, B. J.: Distribution of Deoxyribonucleic acid sequences in fractionated chromatin. Biochemistry 10, 1471—1478 (1971).

Firtel, R.: Personal communication. (1971).

Flamm, W. G., Walker, P. M. B., McCallum, M.: Some properties of the single strands isolated from the DNA of the nuclear satellite of the Mouse. J. molec. Biol. 40, 423—443 (1969).

FLICKINGER, R. A.: The role of gene redundancy and number of cell divisions in embryonic determination. In: RUNNER, M. (Ed.): Changing Synthesis in Development. 29th Growth Symposium pp. 12—39 (1970).

FRASER, M. J., RABIN, E. Z., ALLEN, G.: Purification of *Neurospora crassa* endonuclease-resistant DNA-RNA hybrid. Canad. J. Biochem. 48, 501—509 (1970).

GELDERMAN, A. H., RAKE, A. V., BRITTEN, R. J.: Transcription of Nonrepeated DNA in neonatal and fetal mice. Proc. nat. Acad. Sci. (Wash.) 68, 172—176 (1971).

GEORGIEV, G.: On the structural organization of operon and the regulation of RNA synthesis in animal cells. J. theor. Biol. 25, 473—490 (1969).

GILLESPIE, D., SPIEGELMAN, S.: A quantitative assay for DNA/RNA hybrids with DNA immobilized on a membrane. J. molec. Biol. 12, 829—842 (1965).

HAHN, W. E., CHURCH, R. B.: In: SCHEIDE, VILLIE (Eds.): Cell differentiation. New York: Van Nostrand 1969.

— LAIRD, C. D.: Transcription of nonrepeated DNA in Mouse Brain. Science 173, 158—161 (1971).

KEDES, L. H., BIRNSTIEL, M. L.: Reiterarion and clustering of DNA sequences complementary to histone messenger RNA. Nature (Lond.) 230, 165—169 (1971).

KENNELL, D. E.: Principles and Practices of nucleic acid hybridization. In: DAVIDSON, J. N., COHN, W. E. (Eds.): Progr. Nucl. Acid. Res. Mol. Biol., Vol. 11, pp. 259—302, 1971.

KOHNE, D. E.: Isolation and characterization of bacterial ribosomal RNA cistrons. Biophys. J. 8, 1104—1112 (1968).

— BYERS, M. J.: Studies on the amplification of expressed DNA sequences. Personal Communication (1971).

LAIRD, C. D., McCONAUGHY, B. L., McCARTHY, B. J.: Rate of fixation of nucleotide substitution in evolution. Nature (Lond.) 224, 149—154 (1969).

— Chromatid structure: relationship between DNA content and nucleotide sequence diversity. Chromosoma (Berl.) 32, 378—406 (1971).

McCARTHY, B. J., HOYER, B.: Identity of DNA and diversity of messenger RNA molecules in normal mouse tissues. Proc. nat. Acad. Sci. (Wash.) 52, 915—920 (1964).

— The arrangement of base sequences in DNA. Bacteriol. Rev. 31, 215—258 (1967).

— McCONAUGHY, B. L. Related base sequences in the DNA of simple and complex organisms. I. DNA/DNA duplex formation and the evidence of partially related base sequences in DNA. Biochem. Genetics 2, 37—53 (1968).

— SHEARER, R. W., CHURCH, R. B.: The unstable nuclear RNA of mammalian cells. In: HANLY, E. W. (Ed.): Problems in Biology: RNA in development, pp. 285—312. Univ. of Utah Press 1969.

— CHURCH, R. B.: The specificity of molecular hybridization reactions. Ann. Rev. Biochem. 39, 131—150 (1970).

— Transcription in Mammalian Cells. Address to Symp. on Nucleic Acids, Structure and Function. Banff, Canada, March 1971.

McCONAUGHY, B. L., LAIRD, C. D., McCARTHY, B. J.: Nucleic acid reassociation in formamide. Biochemistry 8, 3289—3295 (1969).

MELLI, M., WHITHFIELD, C., ROA, K., RICHARDSON, M., BISHOP, J. O.: DNA/RNA hybridization in vast DNA excess. Nature (Lond.) 231, 8—12 (1971).

NASS, M. K., BUCK, C. A.: Studies on mitochondrial tRNA from animal cells. II. Hybridization of aminoacyl-tRNA from rabbit liver mitochondria with heavy and light complementary strands of mitochondrial DNA. J. molec. Biol. 54, 187—196 (1970).

NYGAARD, A. P., HALL, B. D.: Formation and properties of RNA-DNA complexes. J. molec. Biol. 9, 125—142 (1964).

PARDUE, M. L., GALL, J. G.: Molecular hybridization of radioactive DNA to the DNA of cytological preparations. Proc. nat. Acad. Sci. (Wash.) 64, 600—606 (1969).

— — Chromosomal localization of mouse satellite DNA. Science 168, 1356—1358 (1970).

PAUL, J.: DNA masking in mammalian chromatin: A molecular mechanism for determination of cell type. In: MOSCONA, A., MONROY, A. (Eds.): Current Topics in Developmental, Biology. Vol. 5, pp. 317—352 Academic Press, New York 1970.

Penman, S., Fan, H., Perlman, S., Rosbach, M., Weinberg, R., Zylber, E.: Distinct RNA synthesis systems of the HeLa cell. Cold Spr. Harb. Symp. quant. Biol. **35**, 561—575 (1970).

Shearer, R. W., McCarthy, B. J.: Evidence for ribonucleic acid molecules restricted to the cell nucleus. Biochemistry **6**, 283—289 (1967).

Sherrer, K., Spohr, G.: Nuclear and cytoplasmic messenger-like RNA and their relation to the active messenger RNA in polyribosomes of HeLa cells. Cold Spr. Harb. Symp. quant. Biol. **35**, 539—554 (1970).

Smith, K.: Personal communication. (1968).

— Church, R. B., McCarthy, B. J.: Template specificity of isolated chromatin. Biochemistry **11**, 4271—4277 (1969).

Walker, P. M. B.: The specificity of molecular hybridization in relation to studies on higher organisms. In: Davidson & Cohn. (Ed.) Progr. Nucl. Acid. Res. Mol. Biol. **9**, 301—342 (1969).

Woodland, H., Graham, C.: RNA synthesis during early development of the mouse. Nature (Lond.) **221**, 327—332 (1969).

Yasmiveb, W., Yunis, J.: Satellite DNA in mouse autosomal heterochromatin. Biochim. Biophys. Res. Commun. **35**, 779—784 (1969).

Kinetic Analysis of the Base Sequence Heterogeneity of RNA Molecules by RNA/DNA Hybridization

Ian Purdom, Robert Williamson, and Max Birnstiel

M.R.C. Epigenetics Research Group, Department of Genetics,
University of Edinburgh, Edinburgh, and
Royal Beatson Memorial Hospital, Glasgow

I. Determination of Base Sequence Heterogeneity in Nucleic Acids

The study of the kinetics of DNA renaturation has yielded much information about the organization of the DNA in eukaryotes into families of nucleotide sequences (Britten and Kohne, 1967, 1968; Wetmur and Davidson, 1968). With a few exceptions the entire assemblage of families of related base sequences of any one organism is treated as a single body of DNA. It is possible to study the renaturation behaviour of a family of genes coding for a specific RNA of genes within the total genomic DNA, but such experiments are tedious to carry out and so far have required a combination of hydroxylapatite fractionation of the renaturing DNA (Flamm et al., 1969) and subsequent identification of the specific DNA segments by RNA/DNA hybridization. By this indirect approach it has been shown, for example, that the 800-fold redundant ribosomal RNA genes of *Xenopus laevis* form a family of closely related base sequence (Grunstein, unpublished). Similarly, after preselection of ribosomal DNA complements from bacteria by RNA/DNA hybridization the renaturation kinetics for these DNA segments indicate that bacterial ribosomal cistrons are similar in base sequence (Kohne, 1968).

Determination of genomic content and/or base sequence heterogeneity of individual families of genes would be greatly simplified if there were a procedure which would allow a kinetic analysis, by analytical RNA/DNA hybridization, of the base sequence content of RNA molecules directly without reference to DNA renaturation data. It is true that, at the outset, such an approach would be restricted to relatively few, well defined RNA species such as rRNA, 5 S RNA and purified messenger RNAs such as 9 S haemoglobin RNA. t-RNA is an especially interesting case since in higher organisms t-RNA genes are highly redundant (Ritossa et al., 1966; Brown and Weber, 1968; Quincey and Wilson, 1969). To what extent are t-RNA molecules related to one another and, do isocoding t-RNA's form single or divergent families of genes?

An analytical RNA/DNA hybridization procedure would probably be the method of choice to elucidate these question. At the moment it would be difficult to study the renaturation behaviour of t-RNA and perhaps also of 5 S RNA genes by measuring the rate of DNA renaturation of the appropriate DNA segments by the technique of hydroxylapatite fractionation and subsequent RNA/DNA hybridization. In order to eliminate interference from DNA lying adjacent to the short t-RNA or 5S RNA specifying DNA stretches, such an approach would require extensive fragmentation of the DNA to segments of the order of 10^4 daltons. Unfortunately our knowledge of DNA renaturation does not extend to this range of low molecular weights.

Another potentially interesting application for analytical RNA/DNA hybridization procedure concerns the base sequence heterogenetiy of ribosomal RNA. In *Xenopus laevis* with only one nucleolar organizer per haploid genome (Kahn, 1962) there is only one family of ribosomal genes (Birnstiel et al., 1969). In cases where the ribosomal genes are distributed over many chromosomes do the ribosomal RNA genes then form several families of genes?

II. Rate of DNA Renaturation is a Measure of Genomic Complexity

The basic principle that lies behind all DNA renaturation experiments is the fact that the rate limiting step in DNA duplex formation is the pairing of complementary base sequences to yield the first "nucleus" of renaturation (Wetmur and Davidson, 1967). The zippering of the complementary DNA strands which follows successful nucleation is very fast and has no, or only little, influence on the rate of the reaction (Wetmur and Davidson, 1967). Since the collision frequency is proportional to the number of molecules per litre it follows that the molar concentration of the molecules, specifically of *each kind* of nucleotide sequence in the solution determines the rate of duplex formation. For this reason the time required for half of the DNA to renature is inversely proportionate to the concentration of the DNA (Britten and Kohne, 1967; 1968), or

$$C_0 \cdot t_{1/2} = c$$

C_0 = initial molar concentration of nucleotides of DNA solutions (Moles/L),

$t_{1/2}$ = time of reaction (sec),

c = constant $\dfrac{\text{(Moles/sec)}}{\text{Litre}}$.

If C_0 is made up from only a few types of nucleotide sequences, as in λ phage DNA, then the rate of renaturation is fast and the constant c small. Where the DNA is heterogeneous because of its large genomic content, as is *E. coli* DNA, the rate of the reaction is slow and the constant large. For these reasons c becomes, under well defined conditions of molecular weight, ionic strength and temperature, a direct measure of the base sequence content of the DNA (Britten and Kohne, 1967, 1968).

Can the same analytical information be extracted from studies of the rate of RNA/DNA hybridization?

III. Rate of RNA/DNA Duplex Formation is a Measure of Complexity of RNA

Nygaard and Hall (1964) observed that when RNA is annealed to DNA in free solution the initial rate of hybrid formation is proportional to the initial concentration of the complementary RNA. Similarly Gillespie and Spiegelman (1965) showed that when RNA is hybridized to immobilized DNA the rate of approach to the plateau is strongly influenced by the concentration of the RNA in solution.

NYGAARD and HALL (1969) also noted that T_2 RNA reacted with T_2 DNA very much more slowly than poly-A reacted with poly-U. They considered this slowness to be a consequence of the requirement of hybridization reaction for bringing together complementary sequences; firstly, the relatively low molar concentration of each complementary RNA in the T_2 system could be the cause for its slow reaction rate and secondly, accidental and ephemeral pairing between quasi-homologous T_2 RNA and DNA segments could temporarily sequester DNA sequences and thus impede rapid hybrid formation. That the first of these considerations is of overriding importance was demonstrated by BISHOP (1969). He showed that there is a direct proportionality between RNA concentrations and the initial rate of hybridization, and that where large amounts of RNA were hybridized with a small standard amount of DNA, duplex formation was a simple function of RNA concentrations. More importantly, when RNA was synthesized *in vitro* on a series of DNA templates of increasing base sequence complexity (λ, T_4 and *E. coli*) and a large, but fixed concentration of this RNA was annealed to a fixed and constant amount of DNA, the rate of hybrid formation was roughly inversely proportional to the complexity of the DNA (BISHOP, 1969).

We have now further investigated this correlation. We here show that within stated limits the mass of the complementary DNA has little or no influence on the time required to reach half of the saturation value, and that over a wide range of RNA the rate of the reaction is independent of the physical molecular weight. We have established optimal conditions for the comparison of absolute rates of hybrid formation and we have correlated the rate of hybridization with the sequence complexity for various RNA species.

A. RNA/DNA Hybridization is Optimal at Relatively High Temperatures

The speed with which RNA/DNA duplexes form is markedly dependent on salt concentration and increases steeply up to about 1 M salt concentration (NYGAARD and HALL, 1964; WALKER, 1969); at higher concentrations still, the rate of the reaction is only slightly augmented. In order to ensure a rapid approach to saturation media containing 0.5 M KCl, 0.01 M Tris, pH 7.2 or 2 × SSC (SSC = 0.15 NaCl; 0.015 M Na_3-citrate, pH 7.2) and temperatures of 60° to 70° were employed. The KCl-Tris medium at 67° was initially evolved to ensure optimal rate of hybridization for T_2 RNA which is of relatively low G + C content (NYGAARD and HALL, 1964). In general, however, such incubation mixtures have become standard for a large variety of RNA preparations for reasons that are pragmatic rather than rational. For the annealing of rRNA, t-RNA and 5 S RNA, which are high G + C RNA species, the standard incubation conditions (2 × SSC or 6 × SSC at 67°) fall far short of those yielding optimal rates and optimal base sequence discrimination. In all these cases the speed of duplex formation can be further increased by raising the temperature. In 6 × SSC the temperatures become so high (see below) that thermal degradation of RNA and loss of DNA from the filters becomes a major problem.

Formamide reduces the dielectric constant of the media and reduces the melting temperature of nucleic acids (MARMUR and Ts'o, 1961). McCONAUGHY, LAIRD and McCARTHY (1969) have successfully used formamide in RNA/DNA hybridization reactions at relatively low temperature without loss of fidelity of base pairing. In this paper we make extensive use of the procedure they have established. In our hands 50% formamide and 6 × SSC represent a good compromise between speed of hybridization and thermal degradation during incubation. We have established the optimal rate temperatures for various RNA preparations in this system.

20 μg DNA from various sources were bound to 13 mm millipore filters (HAWP 0.45 μ white) as described by Birnstiel et al. (1968) and incubated with 1—3 μg homologous RNA/ml for periods short enough not to exceed 20—30% of the hybridization saturation value. In this way we obtained a measure of the initial rates at the various temperatures. The filters were then washed by the batch methods, which is quick and allows simultaneous processing of many filters (Birnstiel et al., 1968) and the disks were counted. Disks with heterologous pro- or eukaryotic RNA were included as background controls.

Fig. 1 shows the optimal curves for the rate of hybrid formation for bacterial ribosomal RNA. The maxima for *B. subtilis* (a) and *Proteus mirabilis* (b) are at 58° and 55° respectively. These temperatures are lower than those of eukaryotic ribosomal

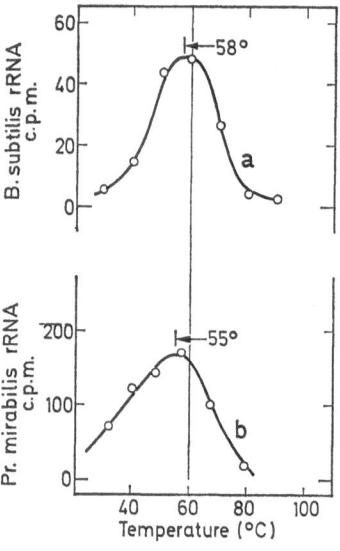

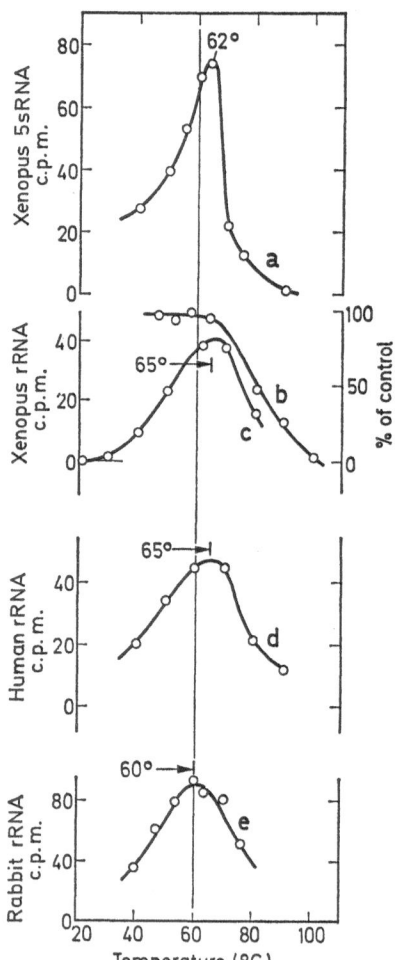

Fig. 1. Determination of optimal rate temperatures in 6 × SSC; 50% formamide for bacterial ribosomal RNA: a) from *B. subtilis* and b from *Proteus mirabilis* (After Birnstiel, Sells and Purdom, 1972)

Fig. 2. Determination of optimal rate temperatures in 6 × SSC; 50% formamide: a for 5S *Xenopus* RNA; b shows the melting profile of *Xenopus* ribosomal RNA/DNA hybrids; c *Xenopus* ribosomal RNA; d human ribosomal RNA; e rabbit ribosomal RNA (After Birnstiel, Sells and Purdom, 1972)

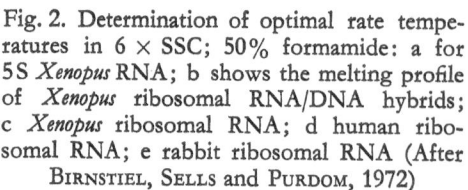

RNA (Fig. 2) which are: 60° for rabbit rRNA (e) and 65° for human (d) and *Xenopus* (c) rRNA. 5S RNA of *Xenopus* (a) reacts fastest at 62° C in 6 × SSC, 50% formamide.

The temperature melting profile (b) for the rRNA/rDNA duplexes are also shown for *Xenopus*. The hybrids for this experiments were manufactured by incubating filter disks in 5 ml of media with 3 μg rRNA/ml for 6 h under standard conditions. The duplexes, bound to the filters were then exposed to different temperature in media identical to the incubation solution (without RNA). The filters were then quickly RNased and washed. It may be seen that at approximately 81° 50% of the rRNA/DNA hybrid is dissociated and rendered sensitive to RNase. This is some 16° higher than the temperature for optimal rate, an increment in good agreement with the original studies of NYGAARD and HALL (1964).

1% formamide reduces the melting temperature by 0.7° (McCONAUGHY et al., 1969). It follows that for optimal rates, rRNA from bacteria and eukaryotes should be incubated at very high temperature in the absence of formamide. The effects of formamide on optimal rate temperatures have been discussed in detail by BIRNSTIEL, SELLS and PURDOM (1972).

B. RNA/DNA Hybridization Curves are Conveniently Presented in a Double Reciprocal Plot

The progress of hybridization is usually presented in a diagram in which counts hybridized or percentage of the DNA hybridized is shown against time. The reaction curve approaches the saturation level asymptotically. Since the saturation-time curve only approximates saturation, but theoretically never reaches it, determination of the endpoint of the reaction is largely a subjective matter, especially at low RNA concentrations.

Approach to saturation, under condition of RNA excess, may also be plotted by recording 1/hybridized RNA or DNA versus 1/time. A straight linear plot for homogeneous or quasi-homogeneous RNA mixtures is then obtained (BISHOP, 1969; BISHOP et al., 1969). Examples of this presentation will be seen later. From this double reciprocal plot it is possible to determine the hybridization saturation value (since the RNA is in large excess) from the intercept at 1/time = 0 (or time = ∞) objectively. In our hands the extrapolation value for some RNA species is about 5—10% higher than the experimental hybridization level obtained after long term incubation of the RNA, where perhaps other factors interfere with the complete attainment of saturation. The theoretical basis for the straight line representing the hybridization reactions is unknown. But because of the great simplicity of the double reciprocal plot it will be used throughout these experiments. In all the experiments described here the RNA species were in excess of DNA and the rates were determined at optimal rate temperatures.

C. The Amount of Complementary DNA Does Not Influence the Relative Rate of Hybrid Formation

Since in the GILLESPIE and SPIEGELMAN technique the hybridization reaction is limited by the nucleation frequency of RNA with complementary DNA (BISHOP, 1969), increasing the amount of DNA by a factor of, say 10, will lead to an increase in the rate of hybrid formation by a factor of 10. That this is so is shown in an experi-

Table 1. Hybridization of rRNA to rDNA in *Xenopus*

Amount of rDNA[a] (μg/disk)	RNA hybridized (cpm)		ratio $\dfrac{\text{cpm}, t = 20}{\text{cpm}, t = \infty}$
	at 20 min	at t = ∞	
0,01	65	200	0.32
0.10	730	2020	0.36
0.33	2031	6940	0.28
1.0	3496	20100	0.17

[a]Disks also contained 20 μg DNA from *Proteus mirabilis*.

ment in which DNA filters containing various amounts of *Xenopus* rDNA (together with 20 μg bacterial DNA) were used (Table 1). The filters were incubated with a large excess of homologous 18S rRNA. In all cases, the RNA concentration was kept constant at 1 μg 18S RNA/ml and even in the case of 1 μg rDNA/disk, the RNA/DNA ratio was > 10.

The absolute initial rate of hybridization (c.p.m. on filter) is roughly proportioanl to the mass of rDNA. If, however, the hybridization is expressed as a fraction of the final saturation level, the relative initial rate of hybridization can be seen to be independent of DNA concentration, at least at low rDNA input. This relationship does not strictly hold at high rDNA concentrations. At 0.33 μg/disk and markedly at 1 μg/disk the reaction slows down and the relative initial rate decreases.

We conclude that at low input the mass of the DNA bound to membranes cannot influence the speed with which saturation is reached. This means, that the degree of reiteration of a particular gene does not determine the *relative* rate of RNA/DNA hybridization. An excess of complementary DNA decreases the relative rate. This slowing down of the reaction probably means that in this case the complementary DNA depletes the RNA solution in the immediate vicinity and/or within the membrane to a measurable extent through rapid hybrid formations, thus effectively reducing the RNA concentration locally and with it the rate of hybrid formation. On this basis one would predict that non-ideal behaviour of the reaction is more pronounced with small disks at high DNA input and with rapidly reacting RNA species.

D. Definition of $C_r \cdot t_{1/2}$

The accurate measurement of initial rates is complicated by two facts: 1) the rate of the reactions slows down measurably even after short intervals of reaction, and this is more pronounced if the approach to saturation is fast. 2) After very short reaction times the level of radioactivity in the hybrid is low. RNA of high specific activity and methods with good discrimination against the background noise are required.

In practice the determination of initial rates is not necessary. Bishop (1969) has demonstrated that the rate of RNA/DNA duplex formation is a simple function of the initial concentration of the RNA (C_r) throughout the course of the reaction. Measurement of the time taken ($t_{1/2}$) to reach half of the saturation value is a more satisfactory

Table 2. Independence of $C_r \cdot t_{1/2}$ from RNA concentration

DNA	Concentration of homologous rRNA (μg/ml)	Complementary rDNA (μg/disk)	$t_{1/2}$ (min)	$C_r \cdot t_{1/2}$ (μg min/ml)
B. subtilis (20 μg)	0.3 (23 + 16S RNA)	0.06	300	90
	1	0.06	75	75
	3	0.06	26	78
	6	0.06	12	72
	9	0.06	8	72

experimental procedure. This is demonstrated in an experiment in which *B. subtilis* rRNA at concentrations of 0.3—9 μg/ml was annealed to a fixed amount of homologous DNA maintaining in all cases a RNA/DNA excess of 10 (Fig. 3). $1/t_{1/2}$, determined at 1/50% saturation = 0.02, $t_{1/2}$ shows a direct proportionality with RNA concentration, as expected. Multiplication of $t_{1/2}$ by the concentration of RNA C_r (expressed as μg/ml or Moles ribonucleotides per litre) yields a product (Table 2) which is within the limits of experimental error, the same for all concentrations.

Evaluation of the hybridization data can be simplified further still. It may be shown that the RNA/DNA reaction can be suitably standardized by expressing the hybridization as percentages of the saturation level, rather than as c.p.m. RNA hybridized. At fixed RNA concentration, well in excess of the complementary DNA, the saturation time-curves and the double reciprocal plots (plotting 1/% saturation versus 1/time) thus become independent of the amount of complementary DNA as long as this amount remains low, as discussed above.

As long as the RNA is in excess, $C_r \cdot t_{1/2}$ is independent of RNA concentration and independent of DNA mass (Table 2). $C_r \cdot t_{1/2}$ varies from one RNA species to another in a way that we shall discuss below.

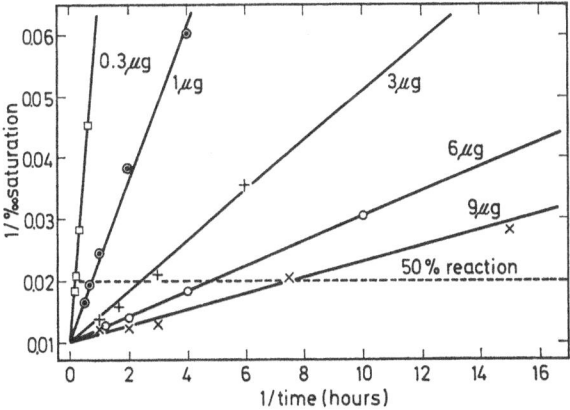

Fig. 3. Double reciprocal plot of the approach to saturation of the ribosomal RNA/DNA hybridisation reaction in *B. subtilis*: $t_{1/2}$ is inversely proportional to RNA concentration (After BIRNSTIEL, SELLS and PURDOM, 1972)

E. $C_r \cdot t_{1/2}$ is Independent of the Size of the RNA

Now it has been shown that the RNA concentration determines the rate of the reaction it is important to study other factors which might be expected to affect the reaction, in particular the effect of the size of the RNA.

When DNA renatures free in solution, the half time is inversely proportional to the square root of the molecular weight of the DNA single strands (Wetmur and Davidson, 1967): although large molecules collide less frequently, the zippering that follows successful nucleation produces more renatured DNA than if the DNA fragment is small. No certain prediction can be made for the RNA/DNA duplex reaction since in this case the complementary DNA is immobilized and also the extent of zippering along RNA/DNA hybrids bound to filters is not known.

We have therefore studied to what extent the physical length of the RNA influences $C_r \cdot t_{1/2}$. To this end 18 S $Xenopus$ rRNA was partially hydrolyzed by exposure to 0.1 N NaOH at 20° for 3, 10 and 30 min and then quickly neutralized by addition of Tris-HCl. The average S values for each rRNA sample were determined as 8 S, 5 S and 3.5 S respectively by sucrose density gradient centrifugation (Fig. 4). The RNA was then incubated at 1 μg/ml under standard conditions of hybridization and the $C_r \cdot t_{1/2}$ determined. From Table 3 it may be seen that in the range of 3.5 to 28 S, i.e.

Table 3

Sample No.	S	cpm hybridized in 20 min
1	28	415
2	8	408
3	5	376
4	3.5	401

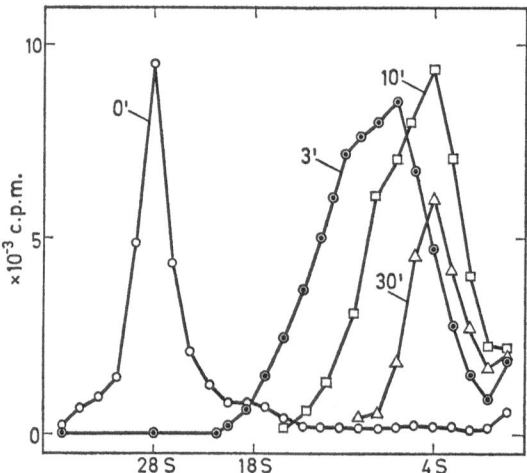

Fig. 4. Sucrose gradient profile of partially hydrolyzed $Xenopus$ 28 S rRNA: RNA was incubated in 0.1 N NaOH at 20° C for the times indicated, neutralized and analyzed on sucrose density gradients

in an (initial) molecular weight range of 2.5×10^4 to 1.5×10^6 the physical length of the RNA has no effect on the progress of the reaction. It may be that partial degradation of the RNA during incubation also minimizes possible effects of the initial molecular weight. Whatever the exact reason RNA/DNA hybridization experiments have the practical advantage that there is no need to determine the size of the RNA fragments used in the reaction in order to measure the rate.

F. Under Optimal Conditions for the Reaction the Secondary Structure of the RNA is Probably Not a Significant Factor in $C_r \cdot t_{1/2}$ Values

Any base or base sequence along a nucleic acid strand is thought to be a potential site for nucleation (WETMUR and DAVIDSON, 1967). RNA molecules such as rRNA and transfer RNA contain a large proportion of nucleotides which are present in paired configuration as pin loops. Such loops make up as much as 60% of ribosomal RNA in prokaryotes, as well as in eukaryotes. If the paired configuration is maintained during the RNA/DNA hybridization reaction, such regions could severely sequester base sequences, and hence reduce the rate of the hybridization reaction. If these regions are more stable than RNA/DNA hybrids, then they also might influence the levels of complementarity between RNA and DNA.

Under optimal conditions for hybrid formation, i.e. high temperature combined with high salt and formamide (see above), most of the secondary structure of ribosomal RNA is lost. This can be deduced from the fact that rRNA at the specified hybridization temperature retains less than 5% hyperchromicity and must be considered to have "melted" to a considerable extent (Fig. 5). The secondary structure of the rRNA would appear to be of little importance under the specified circumstances.

The influence of other factors, such as G + C content of the RNA is not known at the moment. Their influence will have to be determined in suitably chosen model systems.

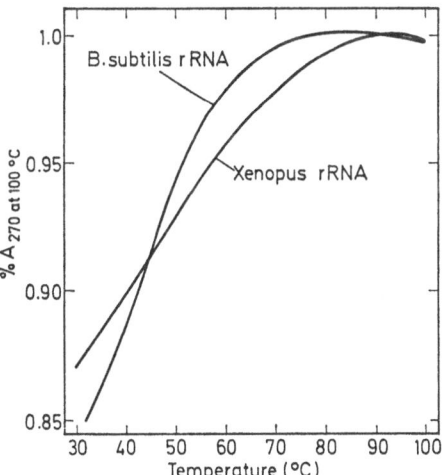

Fig. 5. Melting profile of *B. subtilis* and *Xenopus* rRNA in 6 × SSC; 50% formamide: since formamide adsorbs strongly at 260 mμ, all optimal density measurements were carried out at 270 mμ. At optimal temperatures of 55—58° for bacterial ribosomal RNA and of 60—65° for eukaryotic ribosomal RNA only a small fraction of the hyperchromicity remains

G. $C_r \cdot t_{1/2}$ is a Measure of the Sequence Complexity of the RNA

If an RNA is heterogeneous in base sequence, i. e. is of high genomic complexity, the molar concentration of each kind of base sequence is low and the rate of the reaction must be slow: the rate of hybrid formation at fixed RNA and DNA concentration is inversely proportional to the genomic complexity (see above). For the same reason $C_r \cdot t_{1/2}$ is proportional to the base sequence complexity of RNA and as we have now shown, to it alone, since within certain limits the amount of complementary DNA, size and configuration of the RNA, have no measurable influence. We define the genomic or kinetic complexity of the RNA molecule as the length of RNA base sequence which is unique and non-repeating. For an RNA mixture of (equimolar) molecular species, genomic or kinetic complexity is the sum of the (nonrepeating) base sequences. In both cases it is measured in daltons, as is the molecular weight.

Prokaryotic and eukaryotic 5 S ribosomal RNA is probably the simplest molecular species that can be isolated from a cell with great ease. It possesses a molecular weight of ca. 3.8×10^4 (Williamson, unpublished). Although coded for by a multiplicity of genes (Quincey and Wilson, 1969; Morrell et al., 1967; Zehavi-Willner and Comb, 1966; Tartoff et al., 1970) fingerprint analysis of T_1 digestion products from prokaryotic (Brownlee et al., 1968) and eukaryotic 5S RNA (Williamson, unpublished) show that this RNA is homogeneous in base sequence (Williamson, unpublished; Brownlee et al., 1968) except possibly for minor base substitutions which would go undetected in hybridization experiments.

The cistrons coding for prokaryotic 23 S and 16 S rRNA have been studied by DNA renaturation techniques (Kohne, 1968) and from these experiments it is clear that the multiple ribosomal cistrons of *Proteus mirabilis* are a single family of genes. From this it may be concluded that the *Proteus* rRNA has a base sequence complexity equal to the combined molecular weight of 23 S and 16 S rRNA, i. e. 1.65×10^6 daltons.

When 5S RNA of *Xenopus* or 23S + 16S of *Proteus* is hybridized to their homologous DNA bound to membrane filters it is found that these RNA species have a $C_r \cdot t_{1/2}$ of 0.39×10^{-3} Moles.sec/l and 14.6×10^{-3} Moles.sec/l respectively. *Xenopus* 5S RNA which is 43 times less complex than *Proteus* rRNA anneals 38 times faster. This demonstrates the direct, almost exact interrelationship between complexity of an RNA and its rate of hybridization.

A practical consequence of this is that for a given period of incubation, RNA of low complexity saturates complementary DNA at much lower RNA input. If we consider the case of bacterial rRNA at 3 µg/ml, saturation plateau will be closely approximated after 4 h, since this corresponds to 10 half-lives for the reaction. At the same concentration, 5S RNA has a half-life of only 0.65 min, and incubation for 4 h is excessive and can only augment the contribution of contaminating RNA species!

$C_r \cdot t_{1/2}$ values for other RNA species are listed in Table 4. *E. coli* t-RNA which is made from many different amino acid acceptor species has a $C_r \cdot t_{1/2}$ of 6.8×10^{-3} Moles·sec/l. By comparing this value to that of 5S RNA and bacterial rRNA, we arrive at a complexity of 0.63×10^6 daltons. This is 21 times the complexity of a single t-RNA species.

Xenopus, rabbit and HeLa rRNA all posses a similar base sequence complexity. Once differences in molecular weight have been taken into consideration, it can be deduced that these eukaryotic rRNA species form single families of RNA molecules and have not measurably diverged although encoded by highly redundant genes

Table 4. $C_r \cdot t_{1/2}$ of RNA components

RNA	MW (daltons)	$t_{1/2}$ (min) at 3 µg/ml	$C_r \cdot t_{1/2}$ µg min/ml	$C_r \cdot t_{1/2}$ (Moles sec/L) $\times 10^3$
5S *Xenopus*	3.8×10^4	0.65	1.9	0.39
18S *Xenopus* rRNA	0.7×10^6	8.2	24	4.9
28S *Xenopus* rRNA	1.5×10^6	16	48	9.6
28S + 18S *Xenopus* rRNA	2.2×10^6	26	78	15.6
28S + 18S Rabbit rRNA	2.4×10^6	29	87	17.4
23S + 16S *Proteus mirabilis* rRNA	1.6×10^6	24	72	14.6
23S + 16S *B. subtilis* rRNA	1.6×10^6	24	72	14.6
4S *E. coli* tRNA	3×10^4	11.4	34	6.8

(QUINCEY and WILSON, 1969; BIRNSTIEL et al., 1969; RITTOSSA and SPIEGELMAN, 1965; WALLACE and BIRNSTIEL, 1966). This conclusion is further supported by the high optimal rate temperature and relative thermostability of eukaryotic rRNA/DNA hybrids (BIRNSTIEL et al., in preparation) and is also in agreement with renaturation data on ribosomal DNA from Xenopus (BIRNSTIEL et al., 1969; SPEIRS and BIRNSTIEL, in preparation).

From these preliminary experiments it may be seen that the $C_r \cdot t_{1/2}$ gives a good relative measure of the base sequence diversity of RNA molecules. Once further RNA standards have been studied kinetically it should be possible to extend the present technique further. The kinetic RNA/DNA procedure should then be able to complement fruitfully the well proven techniques of DNA renaturation. Further developments of the kinetic analysis of RNA sequences have been reported elsewhere (BIRNSTIEL et al., 1972).

IV. Summary

Evidence has been summarized which shows that the rate of the RNA/DNA hybridization reaction in the GILLESPIE and SPIEGELMAN procedure is determined by the molar concentration of RNA sequences in the reaction mixture. Further experimentation has established that the time taken to reach half saturation of the complementary DNA ($t_{1/2}$) with RNA in excess is independent of DNA at low concentration and is not measurably modified by the physical length of the RNA sequences in a range of 2.5×10^4 to 1.5×10^6 daltons. Base paired sequences of the rRNA do not interfere with the rate of hybrid formation since under the reaction conditions for optimal rate most of the secondary structure of the RNA is destroyed.

The product of C_r (C_r = molar concentration of ribonucleotides in solution) and $t_{1/2}$ is also independent of RNA concentration and is a constant for each type of RNA. $C_r \cdot t_{1/2}$ can be used as a mesaure of the base sequence complexity of the RNA. The kinetic complexity of various RNA species have been compared to that of 5S rRNA from *Xenopus laevis*. It is shown that in both Xenopus and Rabbit cells ribosomal RNA molecules, synthesized on highly redundant cistrons, possess a base sequence diversity similar to that of bacterial ribosomal RNA.

Acknowledgements

We wish to thank Dr. B. SELLS (Edinburgh) for supplying us with H³-t-RNA from *E. coli*. We gratefully acknowledge many interesting discussions with Dr. J. O. BISHOP (Edinburgh) as well as his gift of C¹⁴-rRNA from *B. subtilis*. We thank Miss SHEENA SMITH for supplying us with *B. subtilis* cells. We are grateful to Miss LINDA PATON for technical assistance.

References

BIRNSTIEL, M. L., SPEIRS, J., PURDOM, I., LOENING, U. E.: Properties and composition of the isolated ribosomal DNA satellite of Xenopus laevis. Nature (Lond.) **219**, 454—463 (1968).
— GRUNSTEIN, M., SPEIRS, J., HENNIG, W.: Family of ribosomal genes of Xenopus laevis. Nature (Lond.) **223**, 1265—1267 (1969).
— SELLS, B., PURDOM, I.: Kinetik complexity of RNA molecules. J. Mol. Biol. **63**, 21—39(1972).
BISHOP, J. O.: The effect of genetic complexity on the timecourse of ribonucleic acid-deoxyribonucleic acid hybridization. Biochem. J. **113**, 805—811 (1969).
— ROBERTSON, F. W., BURNS, J., MELLI, M. L.: Methods for the analysis of deoxyribonucleic acid-ribonucleic acid hybridization data. Biochem. J. **115**, 361—370 (1969).
BRITTEN, R. J., KOHNE, D. E.: Nucleotide sequence repetition in DNA; Carnegie Inst. Year Book **65**, 78—106 (1967). Repeated Sequence in DNA; Science **161**, 529—540 (1968).
BROWN, D. D., WEBER, C. S.: Gene linkage by RNA-DNA hybridization. I. Unique DNA sequences homologous to 4s RNA, 5s RNA and ribosomal RNA. J. molec. Biol. **34**, 661—660 (1968).
BROWNLEE, G. G., SANGER, F., BARREL, B. G.: The sequence of 5s ribosomal ribonucleic acid. J. molec. Biol. **34**, 379—412 (1968).
FLAMM, W. G., WALKER, P. M. B., McCALLUM, M.: Some properties of the single strands isolated from the DNA of the nuclear satellite of the mouse (Mus musculus). J. molec. Biol. **40**, 433—443 (1969).
GILLESPIE, D., SPIEGELMAN, S.: A quantitative assay for DNA-RNA hybrids with DNA immobilized on a membranes. J. molec. Biol. **12**, 829—842 (1965).
KAHN, J.: The nucleolar organizer in the mitotic chromosome complement of Xenopus laevis. Quart. J. micr. Sci. **103**, 407—409 (1962).
KOHNE, D. E.: Isolation and characterization of bacterial ribosomal RNA cristons. Biophys. J. **8**, 1104—1118 (1968).
LOENING, U. E.: Molecular weights of ribosomal RNA in relation to evolution. J. molec. Biol. **38**, 355—365 (1968).
MARMUR, J., Ts'o, P. O. P.: Denaturation of deoxyribonucleic acid by formanide. Biochim. biophys. Acta (Amst.) **51**, 32—36 (1961).
McCONAUGHY, B. L., LAIRD, C. D., McCARTHY, B. J.: Nucleic acid reassociation in Formamide. Biochemistry **8**, 3289 (1969).
MORRELL, P., SMITH, I., DUBNAN, D., MARMUR, J.: Isolation and characterization of low molecular weight ribonucleic acid species from Bacillus subtilis. Biochemistry **6**, 258—265 (1967).
NYGAARD, A. P., HALL, B. D.: Formation and properties of RNA-DNA complexes. J. molec. Biol. **9**, 125—142 (1964).
QUINCEY, R. V., WILSON, S. H.: The utilization of genes for ribosomal RNA, 5s RNA, and transfer RNA in liver cells of adult rats. Proc. nat. Acad. Sci. (Wash.) **64**, 981—988 (1969).
RITOSSA, F., SPIEGELMAN, S.: Localization of DNA complementary to ribosomal RNA in the nucleolus organizer region of Drosophila melanogaster. Proc. nat. Acad. Sci. (Wash.) **53**, 737—745 (1965).
— ATWOOD, K. C., LINSLEY, D. L., SPIEGELMAN, S.: On the chromosomal distribution of DNA complementary to ribosomal and soluble RNA. Natl. Cancer Inst. Monogr. **23**, 449—472 (1966).
SPEIRS, J., BIRNSTIEL, M. L.: in prepatation.
TARTOF, K. D., PERRY, R. P.: The 5 S genes in D. melanogaster, J. Mol. Biol. **51**, 171—183 (1970).
WALKER, P. M. B.: The specificity of molecular hybridization in relation to studies on higher organisms, Progress Nuc 1. Acid. Res. Mol. Biol. **9**, 301—323 (1969).
WALLACE, H., BIRNSTIEL, M. L.: Ribosomal cistrons and the nucleolar organizer. Biochim. biophys. Acta (Amst.) **114**, 296—310 (1966).
WETMUR, J. G., DAVIDSON, N.: Kinetics or renaturation of DNA. J. molec. Biol. **31**, 349—370 (1968).
ZEHAVI-WILLNER, T., COMB, D.: Studies on the relationship between transfer RNA and transfer-like RNA. J. molec. Biol. **16**, 250—254 (1966).

Nucleic Acid Hybridization to Isolated Chromatin

KI-HAN KIM

Department of Biochemistry, Purdue University, Lafayette, Indiana

I. Introduction

Three procedures are in general use for the formation and detection of RNA-DNA hybrids: reactions on a filter (GILLESPIE and SPIEGELMAN, 1965); reaction in liquid (NYGAARD and HALL, 1963); and reaction in agar (McCARTHY and BOLTON, 1964). The inherent limitations of each procedure have been studied in detail (KENNELL and KOTOULAS, 1968). Because of its simplicity, the filter method has been used in various phases of nucleic acid research with good reliability and convenience. Thus it has become possible to estimate the size of that portion of the genome specific for a species of RNA and to measure the appearance of a particular species of RNA during development. However, because of the possibility of nonspecific binding of added RNA to some material, such as basic proteins (which might also render such spuriously bound RNA resistant to RNase), it has been assumed that purity of the DNA fixed on the nitrocellulose filter is imperative.

Since RNA synthesized *in vitro* on isolated chromatin has properties comparable to those of RNA transcribed *in vivo* (BONNER et al., 1963; KENNELL and KOTOULAS, 1968; URSPRUNG et al., 1968; and PAUL and GILMOUR, 1969) the study of isolated chromatin has been encouraged as a means of revealing the nature of chromosomal DNA (see reviews by BONNER et al., 1968; URSPRUNG et al., 1968). Questions relating to the nature of the substances prohibiting RNA polymerase from transcribing the DNA molecule can be experimentally approached more fruitfully if one can distinguish the portions of the DNA molecule which are transcribed from those which are not. This information, in turn, can be related to the functional status of chromosomal DNA.

As a means of answering this type of question, we have examined the feasibility of RNA-chromatin hybridization. This review describes some of the properties of this system and the application of this technique to the study of chromatin.

II. Properties of Isolated Chromatins

Since the whole methodology of RNA-chromatin hybridization is based on the peculiar melting behavior of isolated chromatin, it is appropriate to discuss this subject briefly. When the melting behavior of chromatins prepared by different procedures and from different biological sources was examined, changes in the hyperchromicity of a biphasic nature were found (Fig. 1a, b, and c). Purified DNA's from tadpole liver and pea embryo melt in a well defined manner and exhibit a

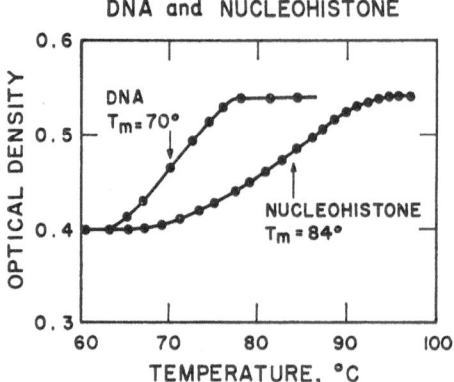

Fig. 1a. Melting profiles of pea embryo chromatin. (BONNER and HUANG, 1964)

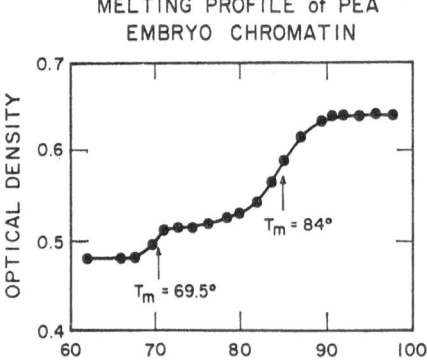

Fig. 1b. Melting profiles of deproteinized DNA and of nucleohistone. (BONNER and HUANG, 1964)

temperature of half melting, T_m, of 77.5° C and 70° C, respectively. Purified chromatins, however, have a two step melting curve. In these cases, the material melting at the lower temperature corresponds to free DNA and represents about 20% of the total chromatin DNA; the remainder, melting at a much higher temperature, apparently represents nucleohistones. Comparative studies of the melting behavior of native chromatin, reconstituted chromatin and deproteinized DNA suggest on the one hand the presence of chromosomal DNA that is not complexed with those classes of histones which stabilize chromosomal DNA. Such a DNA-histone complex has a high T_m and renders DNA inactive as a template for RNA synthesis (BONNER and HUANG, 1964). On the other hand, chromatin melting at a lower temperature does not necessarily represent DNA altogether free from chromosomal proteins. Indeed, it has been shown that the association of histone IV with DNA does not greatly interfere with either melting behavior or template activity (BONNER and HUANG, 1964). However, there is good evidence that the association of other types of histones with DNA is responsible for the increase in T_m of chromatin DNA (BONNER and HUANG, 1964).

The percentage of low melting chromatin in isolated chromatins varies from 0 to 20% depending on the tissue of origin (Table 1). This percentage of low melting

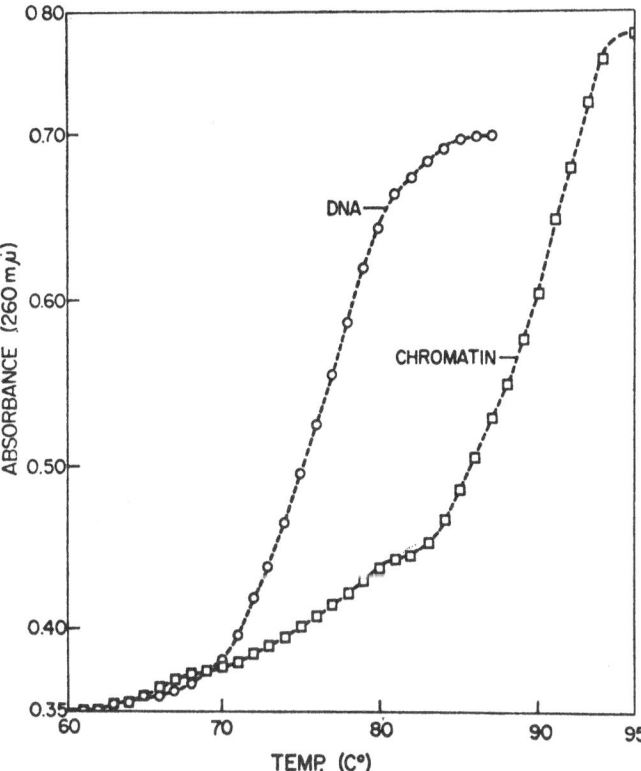

Fig. 1 c. Melting profiles of DNA and of chromatin from tadpole liver. DNA and chromatin were dissolved in a standard saline-citrate buffer (0.015 M NaCl plus 0.0015 M sodium citrate) (KIM and COHEN, unpublished)

chromatin is directly proportional to the effectiveness of chromatin as a template for RNA synthesis. In accordance with the observations on the melting behavior of chromatins and the reports of histone interference with the transcriptional process, PAUL and GILMOUR (1968) and BEKHOR et al. (1969) have shown that only a restricted portion of chromosomal DNA is transcribed by RNA polymerase. It has further been

Table 1. Properties of various chromatins

Source	Lower melting DNA (%)	Effectiveness as a template relative to pure DNA (%)	References
Pea-embryo	20	20	BONNER (1965)
Developing cotyledons	5	10	BONNER (1965)
Duck erythrocytes	0	0.1	BONNER (1965)
Sea urchin sperm[a]	35	2	PAOLETTI and HUANG (1969)
Tadpole liver	20	23	KIM and COHEN (unpublished)

[a] Low melting chromatin had T_m of 66° C whereas DNA had T_m of 47° C.

shown by appropriate competition experiments that the portions available for transcription are different in chromatins from different tissues of the same organism.

Thus, the accumulated literature indicates the heterogeneous nature of chromosomal DNA with respect to their degrees of association with different types of histones, and their efficiency as the template for RNA synthesis. This association of DNA and nuclear protein, in turn, appears to be related to the ease of melting of chromatin. A system has been examined in which RNA is hybridized to chromatin which has been melted at an appropriate temperature, so that only certain portions of the DNA molecule of chromatin can participate in the hybridization. The characteristics of this system are described in this review.

III. Behavior of Chromatin Components during the Hybridization Procedure

Since the only reported studies of RNA-chromatin hybridization were carried out between tadpole liver chromatin and the RNA species induced in tadpoles by thyroxine treatment (KIM, 1967), the behavior of chromatin components during the hybridization procedure will be discussed primarily with this system.

In adapting the hybridization technique of GILLESPIE and SPIEGELMAN (1965) to the RNA-chromatin system, we have asked the following questions: (a) Can chromatin be fixed on nitrocellulose filters? (b) Does chromatin stay on the filters during the hybridization period? (c) Is RNA really annealed to DNA or nonspecifically bound? (d) Does the partially melted chromatin really contain single stranded DNA? (e) Will proteins associated with chromatin leach off during the 60° C incubation in $2 \times SSC$ (0.3 M NaCl plus 0.03 M sodium citrate, pH 7.0) for 12 h? (f) What is the effect of chromosomal proteins? The results of our experiments provide some direct and some suggestive answers to these important questions.

The data summarized in Table 2 show that DNA in chromatin can be quantitatively fixed on a nitrocellulose filter under the standard conditions of GILLESPIE and SPIEGELMAN (1965), and is retained by the filter throughout the hybridization procedure. The behavior of the protein and RNA components of the chromatin during the hybridization period can be seen in Table 3. These data show that, at least quantitatively, the chromatin components suffer very little from the various manipulations involved in the hybridization procedure. In the original experiment of GILLESPIE and

Table 2. Retention of chromatin by nitrocellulose membrane filters

Chromatin treatment	Percent recovery of DNA
Unheated chromatin	100
Chromatin subjected to hybridization conditions	98
Chromatin heated for 10 min at 100° C	100
Chromatin heated and subjected to hybridization conditions	95

Chromatin (0.1 OD at 260 mμ) was fixed on membrane filter by the method of GILLESPIE and SPIEGELMAN (1965). When it was necessary, the fixed chromatin was further subjected to the hybridization conditions. After the treatments, DNA was hydrolyzed from the filter and the recovery of DNA was measured both spectrophotometrically and by the method of BURTON (1956)

Table 3. Retention of chromatin protein and RNA

Treatment of chromatin	Recovery of protein (%)[a]		Recovery of RNA[b] %
	Lysine-³H labeled	Leucine-³H labeled	
Control	100	100	100
Incubated at 60° C for 12 h	96	94	100
Heated chromatin (at 85° C)	90	87	100
Incubated at 60° C for 12 h	94	87	98
Heated chromatin (at 100° C)	94	80	99
Incubated at 60° C for 12 h	93	80	99

[a] Chromatins prepared from tadpoles pulse labeled with lysine-³H or leucine-³H for 3 h were used to determine the retention of protein under various conditions.

[b] Chromatins prepared from tadpoles pulse labeled with uridine-³H for 3 h were used.

Table 4. RNA hybridization to chromatins

Treatment of chromatin	RNA fixed (CPM/filter)
Tadpole chromatin, control	3
Tadpole chromatin, heated	31
Rat liver chromatin, control	0
Rat liver chromatin, heated	2

Tadpole liver chromatin was prepared from animals immersed in thyroxine (2.6×10^{-8} M) for 7 days. d-RNA was prepared from the animals treated with thyroxine for 5 days. Chromatin filters (0.1 OD of DNA) were made as described by GILLESPIE and SPIEGELMAN (1965). Hybridization was carried out in 2 ml of 2 × SSC containing 1950 cpm/120 µg of RNA at 60° for 12 h. Washing procedure was the same as that described by GILLESPIE and SPIEGELMAN

SPIEGELMAN (1965) the "noise" effects of basic proteins were clearly described. The effects of lysozyme and methylated albumin were emphasized. To test the effect of the basic proteins of chromatin, we have examined the non-specific binding of radioactive RNA to chromatin on nitrocellulose filters. As shown in Table 4, unless the chromatins were previously melted, there is no appreciable binding of RNA to chromatin. This suggests that those basic proteins associated with chromatin have no special affinity toward the added RNA. Such non-specific binding would produce a high noise level. It is also shown in the same table that the rat liver chromatin has no appreciable affinity toward tadpole liver RNA, suggesting that the binding between chromatin and RNA is specific. We have also isolated histones from tadpole liver and examined their ability to bind RNA under our conditions. The results of these experiments are shown in Table 5. Regardless of the amount placed on the filter, tadpole liver histones bound only 2 to 3% of input RNA. Most of these bound RNA's were, however, RNase sensitive. The degree of binding of RNA to histone is therefore not as great as that reported by GILLESPIE and SPIEGELMAN (1965) for the binding of RNA to lysozyme. Furthermore, since RNA hybridizes to chromatin and becomes RNase resistant only when it is partially or wholly melted, it appears that RNA is annealed to DNA strands rather than just nonspecifically bound to the components

Table 5. Effect of basic proteins

Exp.	Sample No.	Protein addition	Percentage of input RNA fixed	
			No RNase	RNase
1	1	None	0.2	0.2
	2	20 μg Histones	3	0.4
	3	40 μg Histones	2	0.2
	4	80 μg Histones	2	0.2
2	1	None		0
	2	20 μg Histones		0

In Expt. 1, the mixtures of protein and tadpole liver DNA (50 μg) were fixed and baked as described by GILLESPIE and SPIEGELMAN. The filters with protein and DNA were subjected to hybridization at 60° C for 12 h with 33 μg RNA (1340 cpm). The washing procedures were the same as described by GILLESPIE and SPIEGELMAN (1965) with and without RNase treatment. In Expt. 2, the mixture of 20 μg histones, 40 μg DNA and 33 μg RNA (1340 cpm) in 5 ml of 2 × SSC was incubated at room temperature for 1 h. The mixture was then filtered on the membrane filter and washed as described by GILLESPIE and SPIEGELMAN.

of the chromatin. The fact that DNA strands of chromatin do indeed separate during heating at higher temperatures was demonstrated by the fact that denatured chromatin could be renatured by lowering the temperature slowly.

IV. Application of RNA-Chromatin Hybridization

Tadpole liver chromatin isolated at any stage of thyroxine treatment shows quantitatively the same biphasic melting behavior (Fig. 1c).

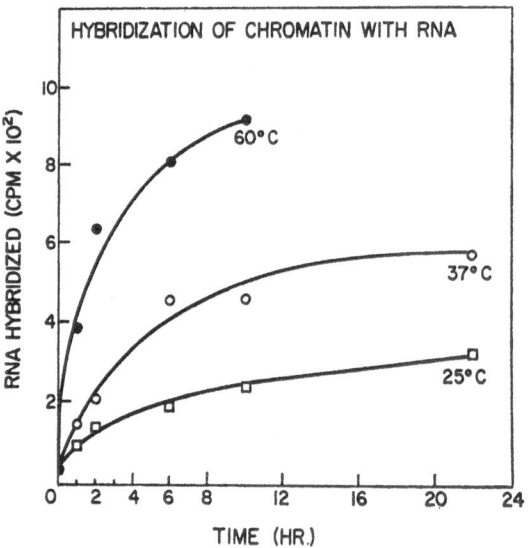

Fig. 2. Time course of DNA-RNA hybrid formation at different temperatures. Partially melted chromatin (2.5 OD at 260 mμ) was fixed on nitrocellulose filter paper and hybridization was carried out in 5 ml of saline-citrate buffer (0.30 M NaCl plus 0.03 M sodium citrate) containing 11 000 cpm (140 μg) of d-RNA. Both chromatin and RNA were prepared from the animals treated with thyroxine for 7 days. (KIM and COHEN, unpublished)

Purified DNA melts in a well defined manner and exhibits a T_m of 77.5° C in dilute saline-citrate buffer. This T_m corresponds to the lower melting chromatin DNA whose T_m was calculated to be 77.5° C. The remainder of the chromatin melts at a T_m of 89.3° C and appears to be nucleohistone. The "free DNA" portion of the chromatin was selectively melted at 85° for 15 min and this partially melted chromatin was fixed on nitrocellulose filters as described by GILLESPIE and SPIEGELMAN. The behavior of the chromatin components during the hybridization period has been previously discussed. Hybridization between fixed chromatin and thyroxine-induced DNA-like RNA (d-RNA) (NAKAGAWA and COHEN, 1967) was also carried out by the method of GILLESPIE and SPIEGELMAN.

Fig. 2 shows the hybridization behavior of partially melted chromatin and d-RNA under different experimental conditions. The maximum hybridization occurred at 60° C for 12 h. Since increasing the temperature had little effect on the extent of hybridization, all subsequent experiments were carried out in 2 × SSC at 60° C for 12 h.

When hybridization experiments were carried out between rapidly labeled RNA and the partially melted chromatins from control and thyroxine-treated animals, the partially melted chromatins from thyroxine-treated animals hybridized RNA 50 to 60% more efficiently than did the chromatin prepared from the control animals (Fig. 3). This difference in the ability to hybridize with RNA is no longer observed when the purified DNA's were compared (Fig. 4). If the low melting DNA portions of the chromatin from treated animals were the only templates responsible for the rapidly labeled RNA, there should be no difference in the degree of hybridization whether chromatin was selectively melted at 85° C or melted completely at 100°. Fig. 5 shows that melting of all the chromatin at 100° C does not increase the extent

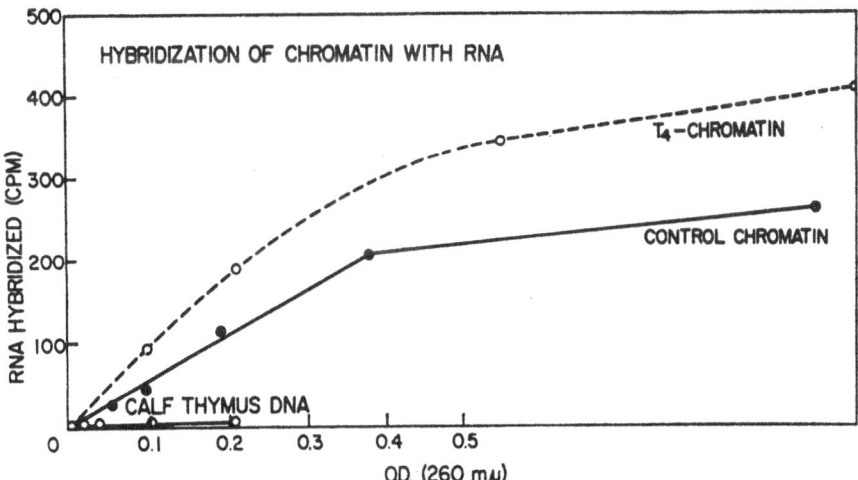

Fig. 3. Partially melted chromatin-RNA hybrid formation. Nitrocellulose filter papers fixed with different amounts (OD$_{260 m\mu}$) of partially melted chromatins (melted at 85°) from control animals and from animals treated with thyroxine (2.6 × 10^{-8} M) for 7 days (T$_4$-chromatin) were incubated with 7900 cpm (17.6 μg) of RNA, isolated from tadpoles treated with thyroxine as described above, at 60° for 12 h. The rest of the procedures were the same as described by GILLESPIE and SPIEGELMAN (1965). (KIM and COHEN, unpublished)

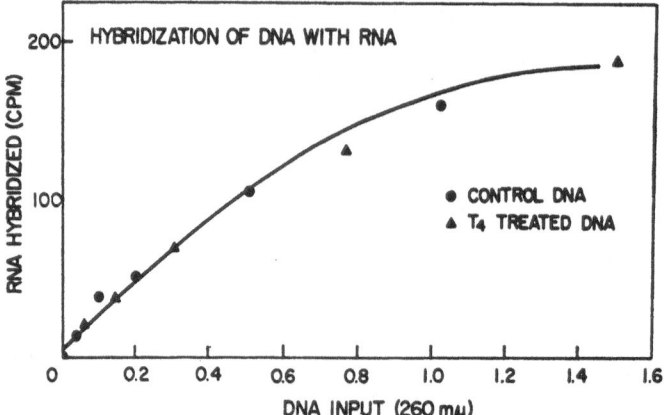

Fig. 4. DNA-RNA hybrid formation at different concentrations of DNA. Nitrocellulose filters with deproteinized and purified DNA's from the control —(●) and thyroxine-treated —(▲) animals were incubated in 5 ml of saline-citrate buffer (0.30 M NaCl plus 0.03 M sodium citrate) containing 9570 cpm (150 µg) of RNA at 60° for 10 h. (KIM and COHEN, unpublished)

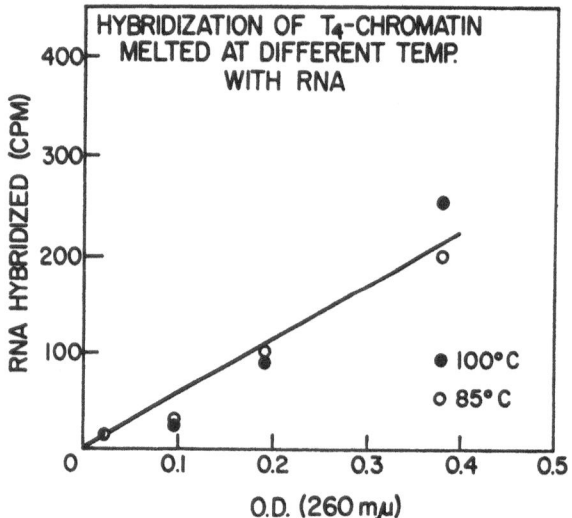

Fig. 5. Hybrid formation between RNA and chromatin (from thyroxine-treated animals) melted at 85° and 100°. Chromatins were melted at 85° and 100° for 15 min. Different amounts of chromatin were fixed on the filter paper and hybridization was carried out in 5 ml of saline-citrate buffer (0.30 M NaCl plus 0.03 M sodium citrate) containing 13 400 cpm (26 µg) RNA. (KIM and COHEN, unpublished)

of hybridization, suggesting that all the hybridizable DNA was already in the form of free DNA and had been melted at 85° C.

When similar experiments were carried out with the chromatin from control animals, totally melted chromatin could hybridize thyroxine-induced RNA to a 90% greater extent than the chromatin partially melted at 85° C (Fig. 6). This suggests that about half of the hybridizable DNA was in the form of high melting nucleo-

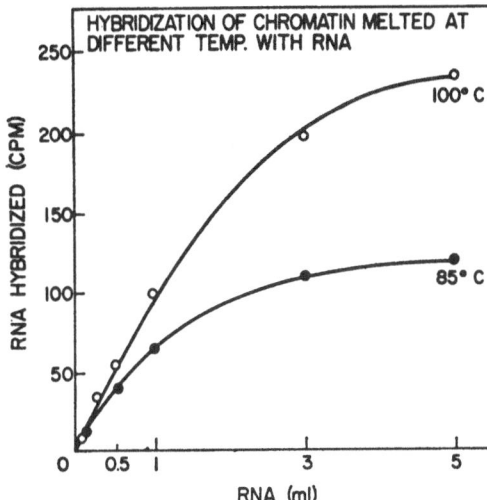

Fig. 6. Hybrid formation between RNA and the control chromatin (from untreated animals) melted at 85° and 100°. Chromatin (0.58 OD at 260 mμ per filter paper) was incubated in 5 ml of saline-citrate buffer (0.30 M NaCl plus 0.03 M sodium citrate) containing different amounts of RNA at 60° for 12 h. Specific activity of RNA was 6500 cpm/33.5 μg/ml. (KIM and COHEN, unpublished)

histones. These studies indicate that the lower melting portion of the chromatin is qualitatively different from the high melting portion and that the difference in hybridization with RNA is due to this qualitative difference. Quantitatively, there is no difference in the relative amount of low melting chromatin or in the chemical composition of chromatins from the livers of control and thyroxine-treated tadpoles (KIM and COHEN, 1966).

V. Concluding Remarks

Although the chromatins fixed on nitrocellulose filters showed little quantitative change during the process of hybridization, subtle qualitative changes such as a possible redistribution of basic proteins or the loss of structurally minor but functionally important components cannot be ruled out at this time.

The presence of single stranded DNA in the melted chromatin is ascertained only indirectly thus far. The attempt to measure this phenomenon directly with the use of ethidium bromide (LEPECQ and PAOLETTI, 1969) was not successful. This might be due to the interference by nucleoproteins. DNAase which is specific for single stranded DNA, such as that described by LEHMAN and NUSSBAUM (1964), might prove to be useful for this study, if more conclusive evidence is required.

The implications of RNA-chromatin hybridization for the interpretation of the mechanism of a hormone action, the regulation of transcription and the biochemistry of chromatin are obvious. The present description of the system is still crude and particular emphasis must be placed on the need for better characterizing the reaction between RNA and chromatin. However, preliminary experiments described here encourage us to pursue the investigation so that our knowledge of chromosomal structure, regulation of transcriptional and developmental processes may be further extended.

Acknowledgements

Sincere appreciation is expressed to Dr. PHILIP P. COHEN for his guidance and direction. Dr. COHEN introduced the author to research in hybridization as a tool in developmental biochemistry and his encouragement and support have been an invaluable asset. I am also indebted to Drs. L. M. BLATT, D. W. KROGMANN, R. L. SOMERVILLE and H. ZALKIN for their helpful criticism of this manuscript.

Preparation of the manuscript was supported by the National Science Foundation Grant GB 8020.

Paper No. 3772 of the Purdue Agricultural Experimental Station.

References

BEKHOR, I., KUNG, G. M., BONNER, J.: Sequence-specific DNA-repressor interaction. J. molec. Biol. **39**, 351—364 (1969).

BONNER, J., DAHMUS, M. E., FAMBROUGH, D., HUANG, R. C., MARUSHIGE, K., TUAN, D. Y. H.: The biology of isolated chromatin. Science **159**, 47—56 (1968).

BURTON, K.: A study of the conditions and mechanism of diphenylamine reaction for the colorimetric estimation of deoxyribonucleic acid. Biochem. J. **62**, 315—323 (1956).

GILLESPIE, D., SPIEGELMAN, S.: A quantitative assay for DNA-RNA hybrids with DNA immobilized on a membrane. J. molec. Biol. **12**, 829—942 (1965).

KENNELL, D., KOTOULAS, A.: Titration of the gene sites on DNA by DNA-RNA hybridization. 1. Problems of measurement. J. molec. Biol. **34**, 71—84 (1968).

KIM, K. H.: Studies on DNA of tadpole liver chromatin. Fed. Proc. **26**, 1603 (1967).

— COHEN, P. P.: Modification of tadpole liver chromatin by thyroxine treatment. Proc. nat. Acad. Sci. (Wash.) **55**, 1251—1255 (1966).

LEHMAN, I. R., NUSSBAUM, A. L.: The deoxyribonucleases of *E. coli*. V. On the specificity of exonuclease I. J. biol. Chem. **239**, 2628—2636 (1964).

LEPECQ, J. B., PAOLETTI, C.: A fluorescent complex between ethidium bromide and nucleic acids. J. molec. Biol. **27**, 87—106 (1967).

McCARTHY, B. J., BOLTON, E. T.: Interaction of complementary RNA and DNA. J. molec. Biol. **8**, 184—200 (1964).

NAKAGAWA, H., COHEN, P. P.: Studies on RNA synthesis in tadpole liver during metamorphosis induced by thyroxine. J. biol. Chem. **242**, 642—649 (1967).

NYGARD, A. P., HALL, B. D.: Formation and properties of RNA-DNA complexes. J. molec. Biol. **9**, 125—142 (1964).

PAOLETTI, R. A., HUANG, R. C.: Characterization of sea urchin sperm chromatin and its basic protein. Biochemistry **8**, 1615—1625 (1969).

PAUL, J., GILMOUR, R. S.: Organ-specific restriction of transcription in mammalian chromatin. J. molec. Biol. **34**, 305—316 (1968).

URSPRUNG, H., SMITH, K. D., SOFER, W. H., SULLIVAN, D. T.: Assay systems for the study of gene function. Science **160**, 1075—1081 (1968).

Hybridization of Nucleic Acids to Chromosomes

Dale M. Steffensen and D. E. Wimber

Department of Botany, University of Illinois, Urbana, Illinois and
Department of Biology, University of Oregon, Eugene, Oregon

I. Introduction

Ever since the method of molecular hybridization ot nucleic acids was conceived, a number of biologists have tried to hybridize labeled RNA or DNA to chromosomes for detection by autoradiography. The first report was by French and Kitzmiller (1967), who hybridized ^{3}H-DNA from *Drosophila melanogaster* to the DNA of the salivary chromosomes of the same species *in situ*. Although their autoradiographs showed radioactivity on the chromosomes well above background, inconsistent results prevented them from publishing a full-length paper. In a second study by Gall and Pardue (1969) there were more complete and convincing findings using the oöcytes of *Xenopus*. In this case, tritium labeled rRNA was found to bind to the rDNA of the nucleolus and to the extrachromosomal rDNA of free nucleoli. The hybridization methods were essentially similar to the nitrocellulose filter method of Gillespie and Spiegelman (1965). The labeled hybrid was unaffected by ribonuclease and unlabeled rRNA competed effectively against labeled rRNA for binding sites. A number of advances have been made in the last two years. We will report some new results, principally on hybridization to chromosomes of the various classes of ^{3}H-RNA, and discuss the prospects of the cytological technique for the future. No attempt is made to review the literature but rather we will evaluate the methods as they can be applied to new and challenging experiments.

Obviously, the major advantage of molecular hybridization to chromosomes *in situ* is that the linear sequence of the genes is maintained. *In theory the location of any specific RNA or DNA hybrid can be determined on a cytological map.* Another advantage of hybridization to chromosomes is that contamination with uninteresting ^{3}H-RNA or ^{3}H-DNA need not always be a problem. Contaminants may bind only to certain known sites in the genome and hence present no difficulty with the analysis of the rest of the genome. This gives a much greater flexibility than is possible by standard hybridization studies using membrane filters.

II. Prerequisites for Hybridization

The detection of gene loci on chromosomes by *in situ* hybridization procedures depends totally upon annealing enough radioactive RNA or DNA at a locus to be able to detect it by autoradiographic means. Several interacting parameters must be considered before one can predict whether an experiment to define a gene location will work, namely, autoradiographic efficiency, hybridization efficiency, cistron redundancy, and specific radioactivity of the RNA or DNA.

A. Autoradiographic Efficiency

Autoradiographic detection of tritium is about 5—10% efficient in cytological preparations (CLEAVER, 1967). It is usually assumed that only one silver grain will be activated by each beta-ray reaching the emulsion and because of the very weak energies involved (mean, 5600 eV), only 10% or less are large enough to be detected autoradiographically. Any method to increase the efficiency, such as using a phosphor coating on the slide, would be desirable in experiments where only very small amounts of radioactivity are bound to chromosomal sites.

B. Hybridization Efficiency

We have made rough estimates of annealing efficiency to cytological preparations by using numbers of autoradiographic silver grains as a measure of binding. About 200 5S RNA cistrons are present in the *Drosophila melanogaster* haploid genome according to TARTOF and PERRY (1970). We usually count between 15 and 30 silver grains over polytene region 56EF, the major site of the 5S cistrons. These numbers are found after a two month exposure period. The specific activity of the 5S^{3}H-RNA was 1.6×10^{11} dpm/μM. A full polytene lateral redundancy is 1024 strands. Thus, the total number of 5S cistrons per polytene chromosome would be about 2×10^5. Using these parameters alone, we estimate a hybridization efficiency of between 3 and 5%. We consider this to be a lower limit estimate of the efficiency, for two reasons. (1) 15 to 30 grains over region 56EF is near saturation of the autoradiograph; hence, the grain count underestimates the total disintegrations. (2) The polytene chromosomes from which we drew our data were often estimated to be below full polytene development, i.e. perhaps only 512 strands. Therefore, it is likely that the actual hybridization efficiency may be 10% or higher.

C. Cistron Redundancy

A substantial portion of the genomes of higher organisms is thought to be highly redundant (BRITTEN and KOHNE, 1968). For example, the cistrons specifying the 18 and 28 S rRNA in the haploid condition are about 130 times redundant in *Drosophila melanogaster* (RITOSSA et al., 1966b) and 450 times redundant in *Xenopus laevis* (BROWN and WEBER, 1968). Individual tRNA cistrons may average about 13 times redundant in *D. melanogaster* according to RITOSSA et al. (1966a). However, redundancies of this magnitude in diploid cells would be difficult or impossible to detect by current hybridization procedures even if extremely radioactive nucleic acid were used. The obvious alternative is to utilize an organism in which the number of cistrons present has been multiplied. The polytene chromosomes of dipterans may have lateral redundancies of 1024 or more. Thus, a cistron having a redundancy of 10 in the haploid state, will be repeated 1×10^4 times or more in polytene chromosomes. The presence of many copies is a great advantage for *in situ* hybridization; if identical cistrons are present in small numbers they obviously cannot be detected by these means.

D. Specific Radioactivity of the RNA or DNA

High specific activity of the nucleic acids is of vital importance for *in situ* hybridizations. Fig. 1 shows specific activity of DNA or RNA plotted against genetic redundancy of a locus in hybridization experiments. The outcome in autoradiographic silver grains per site determines whether or not an experiment is theoretically

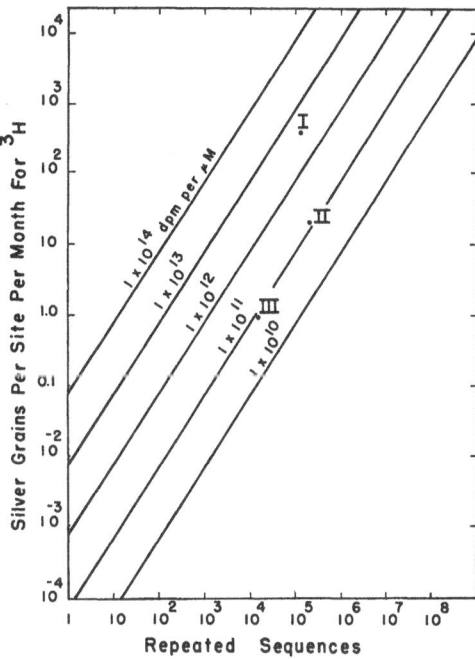

Fig. 1. An estimate of silver grain yield in the emulsion produced by ^{3}H in relation to the number of DNA sequences. The data assume 10% RNA:DNA hybridization efficiency and 10% detection of beta particles by the emulsion. The three experimentally determined values are: I: 4×10^{12} dpm/µM for rRNA, II: 1.6×10^{11} dpm/µM for 5S RNA and III: 9.6×10^{10} dpm/µM for tRNA

possible. We have been successful in producing ^{3}H-RNA with a specific activity of about 4×10^6 dpm/µg from Drosophila larvae grown in a radioactive medium (WIMBER and STEFFENSEN, 1970; STEFFENSEN and WIMBER, 1971). This translates into the specific activities per micromole indicated in Fig. 1. Quite respectable grain numbers should be expected over the nucleolus and the 5S region of polytene chromosomes after one month exposure. On the other hand, only about 1 grain per month would be anticipated over individual tRNA sites, which is a low but a workable level if multiple preparations are examined. Higher specific activities may be obtainable using tissue culture and indeed, PARDUE and GALL (1970) and PARDUE et al. (1970) have routinely produced 5×10^6 dpm/µg RNA and $1-2 \times 10^6$ dpm/µg DNA from *Xenopus* tissue cultures.

In cases where satellite DNA can be separated from main peak DNA, it is possible to use the *E. coli* RNA polymerase system as PARDUE et al. (1970) and MACGREGOR and KEZER (1971) have done to produce highly radioactive RNA (7×10^7 dpm/µg).

From Fig. 1 this latter value would translate into a 1×10^6 daltons of RNA having a specific activity of about 7×10^{13} dpm/μM. Under these circumstances a cistronic repetition of 50—100 would provide high enough grain numbers to be workable. Other methods of labeling the nucleic acids may be of merit. For instance, charging tRNAs with specific radioactive amino acids or enzymatically repairing the 3' Cp CpA with labeled precursors would be useful in determining the sites of production of the particular tRNAs. Catalytic exchange labeling of RNA with tritium might be of merit, provided a high enough specific activity could be produced. Other isotopes may prove to be usable. Some radioactive isotopes of iodine and bromine produce weak enough beta-rays so that fairly good autoradiographic resolution is obtained. Such isotopes, in general, have a much shorter half-life than tritium and hence can have extremely high specific activities. Iodination or bromination of nucleic acids is feasible and a very high specific activity product may be obtained for locating genes that have as few as 10—20 copies.

III. Techniques for Hybridizing Nucleic Acids

Because the techniques of *in situ* hybridization are rather new and still undergoing development, it is often difficult totally to justify certain steps in the annealing procedures. Most of the hybridization conditions have been borrowed from the literature on molecular hybridization where filter annealing practices are used extensively.

A. Fixation

Generally fixation has been in 3:1 (ethanol:acetic acid) in experiments by Gall and Pardue (1969) and Wimber and Steffensen (1970) although in one instance 2:1 (ethanol-chloroform) was used by Buongiorno-Nardelli and Amaldi (1970) and no fixation at all by John et al. (1969). Standard cytological procedures are generally used for the preparations, i.e. squashing in 45% acetic acid and removing the cover glass by the dry ice method. Many tissues can be squashed directly into 45% acetic acid and then post-fixed in 3:1 (ethanol:acetic acid). Some investigators routinely use protein "subbed" slides to prevent the tissue and photographic emulsion from separating from the slide.

B. RNA Removal

The presence of chromosomal RNA could possibly interfere with the binding of radioactive RNA to the DNA of the chromosome. Gall and Pardue (1971) recommend an RNase treatment for removing cellular and chromosomal RNA which otherwise would compete for available hybridization sites. They also suggest a pretreatment with 0.2 HCl to remove any remaining histone.

C. Denaturation

The three most commonly used methods for denaturing the DNA of chromosomes *in situ* are high pH, high temperature, and formamide with elevated temperature. Cytological detail is obscured to some extent by all three methods. For the high pH process, Gall and Pardue (1969) use 0.07 N NaOH at room temperature for two minutes. High temperature methods have been used successfully by John et al. (1969), Jones (1970), and Buongiorno-Nardelli and Amaldi (1970). Five minutes at 100° C in $0.1 \times$ SSC with subsequent quenching at 0° has proved to be effective. The latter method however, may not give more than 1% annealing efficiency.

In our hands, a formamide (90% in 0.1 × SSC) treatment for 2.5 h at 65° gives better banding detail of *Drosophila* polytene chromosomes than the other two methods. Ideally a convenient assay of denaturation should be available in order empirically to find an efficient method for denaturing the DNA of cytological preparations. We have used acridine orange and fluorescence microscopy as an assay for degree of denaturation. Typically RNA and single-stranded DNA fluoresce red, while double-stranded DNA fluoresces green when excited by blue light. Green fluorescence of the chromosomes is usually seen after removal of the chromosomal RNA by RNase treatment according to WOLSTENHOLME (1965). After a 2.5 h treatment of the preparations in 90% formamide at 65° and repeated washing in 2 × SSC at room temperature, an intense red fluorescence of the chromosome is observed. The red fluorescence is reversible by placing unstained preparations into 2 × SSC at 65° for about 20 min; a total return to green fluorescence is seen. We arrived at the 2.5 h formamide denaturation time empirically: treatment for less than 2 h did not give intense red fluorescence, while times longer than 2.5 h impaired the cytological detail. Acridine orange fluorescence was not useful with high pH and high temperature denaturation since only diffuse, light-yellow fluorescence was found.

Recently MACGREGOR and KEZER (1971) reported that a low pH (0.2 N HCl) can be used for denaturing chromosomes. If this procedure is really effective, it would be the method of choice since HCl hydrolysis commonly does not distort cytological detail at the light microscope level. Low pH was the method used by FRENCH and KITZMILLER (1967) in their early experiment.

D. Hybrid Formation

In situ hybrid formation can occur very rapidly. JOHN et al. (1969) have denatured DNA in cytological preparations at 100° C while the chromosomes were bathed in a ³H-RNA solution. Cooling the denatured preparations to 0° in ice still allowed hybrids to form that could be detected autoradiographically. Their usual annealing procedure, however, involved denaturing the DNA on 0.1 × SSC at 100°, cooling and then incubating the preparations from 15 min to 2 hrs at 60° in contact with ³H-nucleic acid in 2 × SSC. In common with Gall's group, we have used reaction times from 4 to 14 or more hours. We are not certain that these long incubation times increase the binding, but at least they seem to do little harm. Quantitative experiments are necessary to answer these uncertainties.

Filter hybridization work suggests that high concentrations of cations (>1 M) interfere with the specificity of the annealing reaction (McCARTHY and CHURCH, 1970), hence concentrations above 6 × SSC are not recommended. Molecular hybridization studies also suggest that relatively high temperatures are best for specificity of the annealing — perhaps slightly below the T_m for purified DNA. Generally *in situ* methods have been successful at 60° or 65°. Lower annealing temperatures (25—40°) could possibly be used if organic solvents such as formamide are employed as considered by McCARTHY and CHURCH (1970). Indeed, lower temperatures would be very useful if ³H-aminoacyl-tRNA is to be hybridized since cleavage of the aminoacyl bond usually occurs at higher temperatures.

Presumably the annealing of the RNA to the complementary strands of DNA is a second order process whose rate is determined by the collision of complementary

4*

strands. With *in situ* hybridization there are two competing processes, the renaturation of the DNA in the chromosomes and the annealing of the RNA to the DNA. In hybridization with filters, complementary DNA strands are immobilized at random and do not reanneal to each other. In chromosomes, complementary strands are in close proximity. Ideally, one would want to devise an immobilizing procedure so that strands cannot reanneal during hybridization. Under present circumstances, a high concentration of RNA in the annealing solution has to be used. Our acridine orange fluorescence studies suggest that the renaturation of a substantial amount of the DNA occurs within the first 20 min after being placed in a salt solution at 65°. Thus, it would seem that the most important period of the hybridization process is the first few minutes after placing the annealing solution on the cytological preparations; stringent precautions should be taken to keep the chemical and physical parameters optimal at that time. John et al. (1969) have used concentrations of 50 µg/ml ³H-RNA, although solutions containing 2 or 3 µg/ml are frequently used by Gall and Pardue (1971). We have used much lower concentrations in some cases, i.e. 0.2 µg/ml and Gall and Pardue (1971) report that with highly radioactive RNA produced by the *E. coli* polymerase system, concentrations as low as 0.01 µg/ml can be used successfully.

Typically, 0.1—0.2 ml of ³H-nucleic acid solution is placed on the denatured preparation, a cover glass added and the slide placed in a pre-heated plastic petri dish *on glass rods*. The petri dish contains several ml of the same salt solution The petri dish is then maintained at the proper annealing temperature for the duration of the experiment.

E. Removal of Unbound RNA

Following the annealing procedure, the cover glass is floated off in $2 \times$ SSC and the preparation is washed copiously in $2 \times$ SSC. This is followed by a 1 hr treatment with RNase to remove any non-hybridized RNA (Wimber and Steffensen, 1970; Gall and Pardue, 1971). In DNA:DNA hybrids, several saline washes at 55—60° remove the non-specifically trapped DNA (Pardue and Gall, 1970; Jones, 1970).

IV. Unresolved Methodical Problems

The procedures for the *in situ* hybridization to chromosomes are so new and exploratory that the techniques are qualitative and still evolving. Eventually the method must be made quantitative. One of the main problems remains the complete denaturation of DNA without losing the integrity of the chromosomes or cytological detail. Most of the fixation procedures discussed in the previous section are variations of standard chromosome methods and are known to remove chromosome protein. The fixation in 45% acetic acid removes most of the basic protein (Dick and Johns, 1967), while 0.2 normal HCl will remove most of the remaining histones. Either of these latter treatments will leave most of the acidic proteins intact. The denaturation in 0.07 N NaOH used by Gall's group must remove some acidic proteins (Hnilica et al., 1965). However, none of the fixation procedures and denaturation methods remove all of the proteins from the chromosomes. We have shown this to be true by the following observations. Salivary gland chromosomes, fixed or treated by one or several of the preceding methods, when placed in a buffered pronase solution show dissolution within a few minutes. After 10—15 min the physical integrity of the chromosomes is destroyed. The DNA can be seen as "going into solution" if acridine orange is used to stain the chromosomes before the pronase treatment and the preparation observed by fluorescent microscopy. In this case the DNA is in a diffuse state and no longer associated with any recognizable chromosome after 10 to 15 min exposure to pronase.

Almost all of the hybridization procedures discussed so far are a compromise between making the DNA strands available and leaving some protein and structure present for reasonably good cytology. At present nothing is known quantitatively about the type and amounts of steric hindrance contributed by the various chromosomal proteins in the hybridization procedure. Future quantitative studies aiming for high hybridization efficiency will require the solution of this question.

As previously mentioned, complementary DNA strands reanneal to each other during incubation in the solution of $2 \times SSC$ in the chromosomes. These melted complementary strands must be very close together. Nothing is known about the configurations they take when denatured, nor is anything known about the constraints or lack of constraints in cytological preparations. Ideally, a new or revised method should be devised to immobilize single stranded DNA so that self-annealing becomes negligible. Until these problems are resolved in a predictable way, precise quantitative hybridization *in situ* is not possible.

V. Precautions against Artifacts

It is certainly possible to obtain an autoradiograph in nucleic acid hybridization experiments with chromosomes. But the autoradiograph is not a stringent proof that complementary base pairing of DNA to DNA or RNA to DNA has actually occurred. Aside from post-treatment with enzymes that degrade single-stranded molecules, one test of hybridization is to determine the thermal stability, T_m, of the new hybrid. Non-specific or partial binding will be labile already at slightly elevated temperatures, while complementary strands will require a temperature approaching the T_m of the original DNA to dissociate the complex. LAIRD and MCCARTHY (1968a, b) compared the stability of DNA:DNA and rRNA:rDNA hybrids by this method, using nitrocellulose filters. Cytological preparation with putative molecular hybrids could be subjected to the same procedure. Molecular hybrids would be stable to nearly the T_m of the chromosomal DNA. Any radioactivity that remained after complete melting had occurred would be artifactual.

Competition experiments with unlabeled nucleic acids are another way of testing the reality of a molecular hybrid. Complementary strands will compete stoichiometrically with labeled strands. Closely related strands will compete in proportion to their related sequences, while unrelated sequences will not compete at all. Unlabeled tRNA was used by STEFFENSEN and WIMBER (1971) to compete with ³H-tRNA and the discrete labeling at tRNA genes was obliterated. Also a 100-fold excess of cold 5S RNA removed 97% of the radioactivity in competition with ³H-5S RNA. Using this procedure, the genes coding for 5S RNA in *Drosophila melanogaster* have been located at the bands 56EF on the right arm of chromosome 2 as shown in Fig. 2 and discussed in more detail in Section VII.

Another source of artifacts is the use of labeled nucleic acids of low purity. For example, phenol purification of RNA and one passage through a column may not

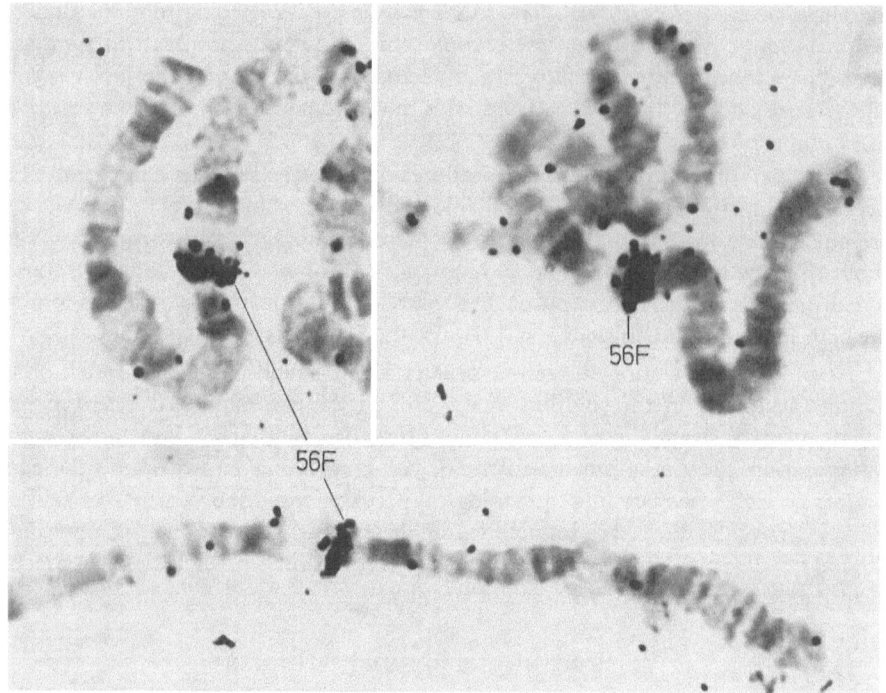

Fig. 2. Autoradiographs of salivary chromosomes of *Drosophila melanogaster* hybridized with ³H-5 S RNA. The radioactivity locates the genes for 5 S RNA at 56 EF on 2 R

suffice. In one experiment ³²P labeled 4 and 5 S RNA was taken from a MAK column. When we used this RNA for autoradiographic hybridization to *Drosophila* salivary chromosomes, autoradiographs of the preparation showed that fine particles from the column gave "³²P stars" distributed at random on the slide. Such radioactivity would be recorded as "noise" by the nitrocellulose filter method too, unless further purification steps were taken. Low molecular weight RNA can be purified on a benzoylated DEAE cellulose column, where messenger RNA and rRNA are retained while tRNA and 5 S RNA are eluted (GILLAM et al., 1967). Preparative gel electrophoresis is also effective, since tRNA and 5 S RNA are widely separated on 9% acrylamide gels, as are 18 S and 28 S RNA on 2.8% gels.

VI. Repeated DNA Sequences

Repeated or redundant sequences offer the most promising prospects for nucleic acid hybridization to chromosomes. The best understood redundant DNA so far is the rDNA contained in the nucleolar organizer that codes for rRNA. In *Drosophila*, RITOSSA and SPIEGELMAN (1965) have found that rDNA accounts for about 0.3% of the genome. Most organisms probably contain the same proportion of rDNA. The major exceptions are the stage dependent changes in oögenesis found in Amphibia and other organisms, where the rDNA is greatly amplified by means of extra free nucleoli (for references see GALL, 1969; MILLER and BEATTY, 1969). The electron micrographs of the peripheral nucleoli of *Triturus* by MILLER and BEATTY (1969) are superb illustrations of their serial duplication and synthetic properties.

As expected, the most rapid advances using the technique of *in situ* hybridization have been with the highly repeated DNA sequences. The nucleolar organizer DNA (rDNA) was first visualized by nucleic acid hybridization in *Xenopus*, by GALL and PARDUE (1969), and after their work the same type of experiments were carried out by others in a variety of organisms. The procedure was to isolate highly labeled ^{3}H-rRNA and use it to anneal with the rDNA at the nucleolus organizer on chromosomes. Another method was to separate rDNA as a satellite from main band DNA on a cesium chloride gradient. Here the high G + C satellite is primarily the rDNA. This rDNA is then used with the *E. coli* RNA polymerase to make highly radioactive rRNA for hybridization. Aside from the basic information and perfection of the technique, one of the most interesting features is that *Xenopus* rRNA hybridizes with rDNA of a number or organisms including *Drosophila melanogaster*. The latter seems extraordinary since the rDNA of *Drosophila* does not have a high G + C content. From these experiments one concludes that homologous segments must exist between two species as wide apart as an amphibian and an insect, and have been preserved in evolution. Drosophila is one of those few organisms that have high AT rDNA; the majority of organisms from bacteria through higher vertebrates have rDNA with about 60% GC.

The mouse has a large amount of satellite DNA with a high A + T content. These highly repetitive sequences have been characterized by BRITTEN and KOHNE (1966, 1968) and their findings showed satellite DNA to comprise nearly 10% of the total DNA of the mouse genome. In another study, YASMINEH and YUNIS (1969) presented evidence that the mouse satellite DNA was heterochromatin. In their study they separated euchromatin from heterochromatin. On a cesium chloride gradient, the heterochromatin fraction from males was much enriched for the satellite DNA. Soon thereafter, PARDUE and GALL (1970) and JONES (1970) isolated this high A + T satellite, made complementary RNA from the DNA by the *E. coli* RNA-polymerase system and hybridized the labeled RNA to mouse chromosomes. The autoradiographs show the label to be very near or contiguous to the centromere in every chromosome except the Y. It would appear, therefore, that this highly repetitive DNA is chromocentric DNA. A similar study was done by RAE (1970) where the DNA of *Drosophila* was separated by cesium chloride, and a fraction selected from the gradient on the shoulder of the main band. This DNA was then used in the *E. coli* RNA polymerase system to make labeled complementary RNA to hybridize back to *Drosophila* chromosomes. The label appeared primarily at the chromocenter, the region of joined centromeres. ECKARDT (1970) has reported a similar experiment in another dipteran fly, *Rhynchosciara*. Here the main band DNA has a buoyant density of 1.695 g/cm³ and the satellite has a density of 1.681 g/cm³. These repetitive sequences in the satellite DNA are not only at the centromere but also at the very tips of the chromosomes, the telomeres. It appears from these very recent experiments with repetitive DNA that much new information can be gained from locating repeated sequences in the chromosomes of a variety of organisms. The analysis of this DNA, which for years has gone under the term heterochromatin, will provide substantial insight into its role in gene regulation, position effect, late replication, and chromosome pairing and movement. One can hope to find a chemical basis for partitioning

groups of phenomena in relation to a particular species of DNA and to show their
location on the chromosome.

According to Britten and Kohne (1966; 1968) all higher organisms examined
from *Euglena* to man possess repetitious DNA. If we exclude the rDNA and the mouse
satellite, then 15% or more of the remaining genome is repeated with 1000 to perhaps
100000 copies per haploid complement. At present there are no assigned functions
or clues as to the chromosomal localization of these repeated sequences.

Following the work of these authors, DNA can be separated into three fractions
— the highly repetitious DNA, the repetitious DNA, and the unique or non-repeated
DNA. In the annealing reaction the highly repeated sequences become double-
stranded quite rapidly and can be recovered selectively. Double-stranded DNA binds
to hydroxyapatite columns and can be eluted by increased temperature or higher salt
concentration. One is bound to recognize these different fractions readily because of
their vastly different rates of reassociation. These rates are usually plotted as the
function, C_0t, defined as the product of concentration and time of incubation (moles
nucleotide $\times$ sec per l). The conditions for hybridizing any particular DNA fraction
must be chosen carefully. Before attempting any experiments with chromosomes,
it would be wise to establish the conditions by these standard biochemical methods.

According to Britten and Kohne (1966, 1968), the non-repeated or unique
sequences of most higher organisms amount to about 70% of the genome. It is
feasible to hybridize nucleic acids to these genes, using polytene chromosomes. In
our recent study (Steffensen and Wimber, 1971) we were able to detect tRNA genes
that are serially repeated, amounting to no more than 4×10^5 Daltons of DNA.
Unique sequences, like genes making messenger RNA (mRNA), might be near this
molecular weight and amenable to study. The major difficulty will be to isolate
enough specific mRNA for hybridization. High concentrations of mRNA may be
necessary if reannealing of DNA strands takes place during long hybridization periods.

One way to attack the problem of mRNA hybridization would be to isolate
polysomes from various stages of development. For example, we might choose
Drosophila larvae and isolate the salivary gland polysomes at different times during
larval development. However, it will be difficult to isolate the ^{3}H-RNA for hybridization and the experiment will require new microchemical techniques. The bands
which puff and those which are labeled with ^{3}H-uridine are well known in *Drosophila*;
thus one could determine immediately which gene provides the RNA that moves in
to the cytoplasm as messenger in salivary gland cells. Such experiments could be
made more sophisticated by isolating microsome fractions on sucrose gradients apart
from free polysomes. The four or five different sub-fractions of the polysomes might
provide future information about coordinated protein synthesis as partitioned in the
cytoplasm.

One might expect the information from the hybridization with cytoplasmic RNA
from either *Drosophila* or *Chironomus* to be very revealing with regard to the function
of chromosome puffs. The puffs in *Drosophila melanogaster* have been extensively
characterized by Becker and Schultz (cited by Lindsley and Grell, 1967) and
Ashburner (1969a, b). *Chironomus tentans* is another ideal organism for study since
puffing has been examined very intensively and the puffs which synthesize RNA have
been mapped by Pelling (1964, 1970) using autoradiographic and cytochemical
techniques. Puffed regions have been assumed to have the function of synthesizing

and transporting RNA to the cytoplasm. If these predictions are correct, one might expect that messenger RNA in the cytoplasm would hybridize preferentially with puffed regions at any particular stage during development. By the same token unpuffed loci which are making RNA at low levels would not provide RNA in the cytoplasm and would not have representative sequences in the cytoplasm. All of this provides a way to attack the problem of nuclear RNA in *Drosophila*. In HeLa cells it is postulated that 90% of nuclear RNA is destroyed or turned over while only 10% gets to the cytoplasm (DARNELL et al., 1970).

VII. Location of the Genes Coding for tRNA and 5S RNA in Drosophila

A. 5S RNA Genes

Various experiments have been designed to test if 5S RNA was coded in the nucleolus organizer (see TARTOF and PERRY, 1970). Everyone expected the genes for

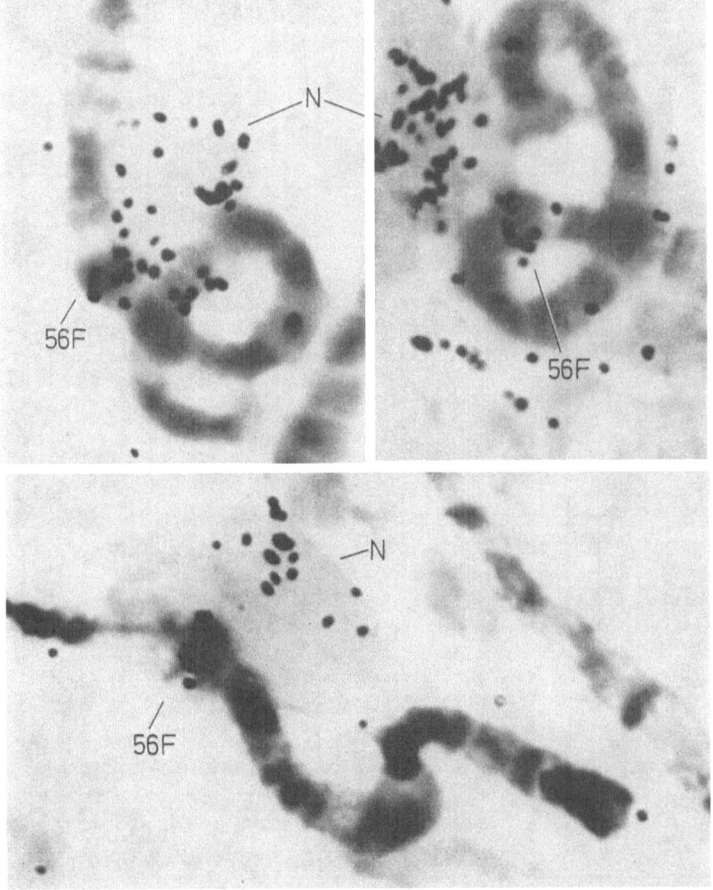

Fig. 3. Nucleoli (N) attached to 56 F of *Drosophila* salivary chromosomes. The nucleoli are radioactive and are so identified by the ^{3}H-rRNA contamination present in the [3]H-5S RNA fraction used in hybridization

Table 1. Location of sections labeled with ^{3}H-tRNA correlated with the location of Minute loci and puffs in third instar larvae and prepupae in *Drosophila melanogaster* salivary chromosomes

X-chromosome				Chromosome 2							
				2L (sections 21 to 32B)				2R (sections 41 to 60)			
^{3}H-tRNA[a] labeled sections	Puffs	Location of min gene	Location of min bands	^{3}H-tRNA[a] labeled sections	Puffs	Location of min gene	Location of min bands	^{3}H-tRNA[a] labeled sections	Puffs	Location of min gene	Location of min bands
1B	1B	M(1)Bld	(1C)	21A				41A—F		M(2)S2	(41A)
2A				21C	21C	M(2)21C1-2	(21C)	42A	42A		
2B	2B			21D	21D			43B	42B	M(2)38b	(42A—44D)
	2E				21E			43B	43E		
	2F			22A	21F				44A		
3A	3A			22A	22A				44B		
	3B				22B				44E		
	3C			22C	22C			45A			
3D	3D				22D				46A	M(2)40c	(45E—47F)
	3E	M(1)3E	(3E)	23B	23B				46F		
		M(1)4BC	(4BC)		23C				47A		
4E	4E				23D				47B		
	4F				23E				47C		
5C				24F		M(2)Z	(24E—25A)	47F			
5D		M(1)30	(5E—7B)	25A	25A				48A		
6E					25B				48B		
	7A			25C	25C	M(2)S1	(25A—25E)		48D		
	7B				25D			49B			
7D	7D			26B	26B				49E		
	8B			27B					49F		
	8D			27C	27C			50A			
8E	8E				27F			50C	50C	M(2)d	(49E—52B)
	8F				28A			50D	50D		
	9B			28C					51D		
	9E			28D	28D				51E		
	9F			29C							
				29D							

Sect. (10–18)	M(1) locus	Sect. (29–32)	Sect. (52–60)	Sect. (52–60)	M(2) locus
10A		29F	52A	52F	
10E	M(1)k (9E—10B)	30A	52C	53A	
10F			52D	53F	
11B		30A	52E	54E	M(2)S7 (52E—53B)
11C		30C	53F	55D	
12A		30D	54A	57C	M(2)b (53F—56E)
12E		32A	55B	57F	M(2)173 (55D—57A)
13A			55E	58D	
13B			56D	60A	
13C		M(2)e (31A—32F)	56E		
13E			58B		
13F			58C		
14B			58D		
14D			58E		
14E			58F		M(2)1 (58F)
14F			59D		
15A			59F		
15B			60A		
15C			60B		
16A	M(1)$_0$Sp^a (16A)		60C		
16B			60D		M(2)c (60E)
16C					
16D					
16E					
16F					
17A					
18A					
18D	M(1)n (18D)				

[a] Only the sections exhibiting highly significant labeling (P <0.001 by Poisson distribution test) are given in one or both the 4S and 5S RNA fractions containing tRNA (Steffensen and Wimber, 1971). Sections, where labeling was less significant (<0.01 or 0.05) are not listed. The cytogenetic localization of Minutes was obtained from Lindsley and Grell (1967), as were some of the puff locations. Most of the information on puffing sites was taken from Ashburner (1969a, b).

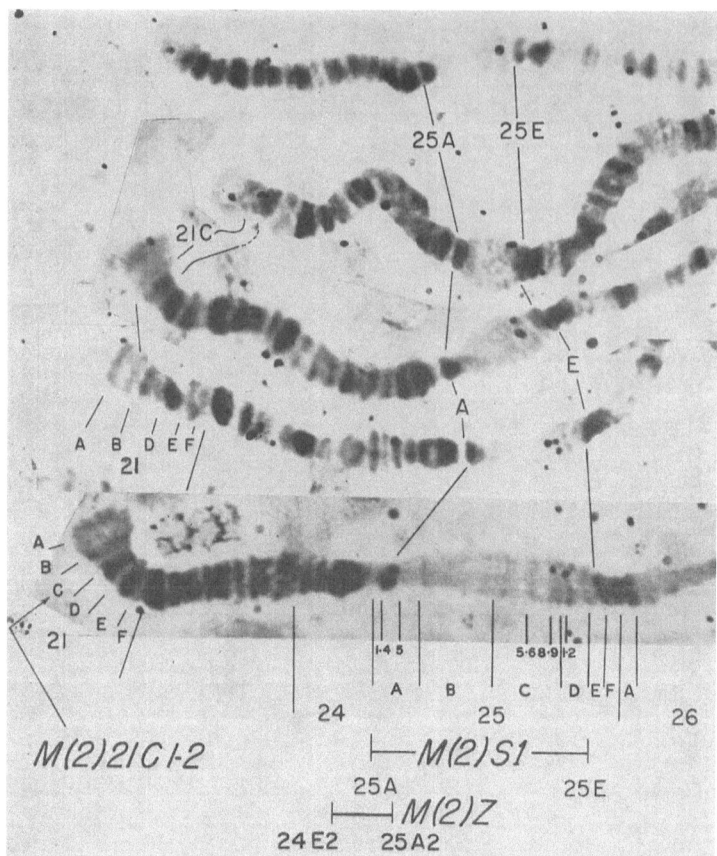

Fig. 4. Autoradiographs of the left arm of chromosomes 2 (2L) in *Drosophila*. The radio-activity in the puff at 25 C 8.9 is attributed to the binding of ³H-tRNA and is considered to be one of the genes coding for tRNA. Three Minute (M) loci on the tip of 2L are located according to Lindsley and Grell (1967)

5 S RNA to be within or at least adjacent to the nucleolus organizer on the X-chromosome. We know now that the 5 S RNA genes are not located on the X-chromosome, but on chromosome 2 at region 56 EF (Wimber and Steffensen, 1970). The calculations and additional studies presented in section III suggest that all of the genes coding for 5 S RNA are located at region 56 EF. No mutants are known to map at this region of the second chromosome (Lindsley and Grell, 1967).

How is 5 S RNA transferred from region 56 EF to the nucleolus for assembly into 60 S ribosomal particles? We have only one clue at the moment. An occasional cell in autoradiographs shows nucleoli or pseudonucleoli attached directly to 56 F (Fig. 3). We are certain that these associations are nucleoli since the 5 S RNA fractions employed in hybridization were contaminated with ³H-rRNA that hybridized to the rDNA in the nucleolus. Competition with unlabeled rRNA removed nucleolar radioactivity to a low level but was without effect on ³H-5 S RNA binding to 56 EF bands. The immediate and obvious interpretation is that the nucleolus is picking up newly synthesized 5S RNA from 56EF for the assembly of 60S particles. This latter assumption, however, turned out to be wrong or at least too simple. In excised salivary glands labeled with ^{3}H-uridine, the 56F bands were never labeled from mid-third instar up to prepupal stages in studies by HOLMQUIST (unpublished). In those cells where 56F was associated with the nucleolus, 56F was turned off and not making RNA at the time one might expect. During all of the developmental stages examined, the nucleolus was always synthesizing rRNA at a rapid rate. Ribosomes must be maturing, which requires new 5S RNA. At this time we have no answer to the dilemma. We are convinced that interchromosomal connections and associations with the nucleolus to certain chromosome sites are phenomena of significance. In many ways the nucleolar connectives are analogous to ectopic pairing, summarized by KAUFMANN and IDDLES (1965). In any case the method of *in situ* hybridization has provided the means and the insight to examine a classical observation of intranuclear recognition.

B. Transfer RNA

A year ago we thought it would be nearly impossible to locate the genes coding for tRNA. We tried anyway and the experiments worked.

Table 1 shows cytological map locations for the ^{3}H-tRNA:DNA hybrid sites on the X-chromosome and for chromosome 2, taken from STEFFENSEN and WIMBER (1971). The positions of Minute loci are also given in Table 1 using the cytogenetic information found in LINDSLEY and GRELL (1967). The Minute mutants are listed in view of K. C. ATWOOD's hypothesis that Minute loci are tRNA genes (RITOSSA, ATWOOD, and SPIEGELMAN, 1966a). The correlation is reasonably good but there are enough examples of non-correspondence to withhold final judgement. Further study improves the statistical aspect of the data, besides providing internal controls like Minute deficiency heterozygotes and chromosomal aberrations, where the wild type allele of Minutes at the break point is moved to a new position in chromosomes used for hybridization.

Another feature in Table 1 is that many of the putative tRNA genes are located in sections which puff. It would be interesting to know if the correlation has developmental significance. Are certain tRNA species being synthesized, while others are not? Could this differential synthesis have an effect on the type of proteins being made in the cytoplasm by regulating translation? Once the remainder of the genome has been analyzed, we will be in a better position to answer how many genes code for tRNA. At present we estimate there will be some 130 to 140 tRNA genes for the whole genome.

Another area of future study will be to map the tRNA genes more precisely. The prospects are illustrated in Fig. 4. In good cytological preparations the radioactivity can be localized to a band or two, given repeated scoring. As illustrated in Fig. 4, the bands 25C 8—9—10 are radioactive in almost every salivary gland cell when hybridized with ^{3}H-tRNA. If we assume an average repetition of some thirteen cistrons for this tRNA gene, then there would be less than a haploid value of 10^6 Daltons of serially repeated DNA at 25C 8—9—10. This amount is much smaller linearly than the molecular weight of DNA in a band. Thus, one would predict that high resolution autoradiography will locate this tRNA gene within a single band.

What are the functions of the remaining DNA in such bands and the "tRNA puffs" in general? It is tempting to speculate that some remaining genes might have enzymatic functions to alter the tRNA after transcription.

Acknowledgements

We wish to thank Dr. Gerald Holmquist for critical comments and Mrs. Lilian Cooper and Mrs. Doris Wimber for their technical assistance. Research support was given by NSF GB-18922 and NIH grants GM 8465 and GM 11702.

References

Ashburner, M.: Patterns of puffing activity in the salivary gland chromosomes of Drosophila. II. The X-chromosome puffing patterns of *D. melanogaster* and *D. simulans*. Chromosoma (Berl.) **27**, 47—63 (1969a). IV. Variability of puffing patterns. Chromosoma (Berl.) **27**, 156—177 (1969b).

Britten, R. M., Kohne, D. E.: Nucleotide sequence repetition. Carnegie Inst. Wash. Year Book **65**, 87—106 (1966).

Britten, R. J., Kohne, D. E.: Repeated sequences in DNA. Science **161**, 529—540 (1968).

Brown, D. D., Weber, C. S.: Gene linkage by RNA-DNA hybridization. I. Unique DNA sequences homologous to 4 S, 5 S RNA and ribosomal RNA. J. molec. Biol. **34**, 661—680 (1968).

Buongiorno-Nardelli, M., Amaldi, F.: Autoradiographic detection of molecular hybrids between rRNA and DNA in tissue sections. Nature (Lond.) **225**, 946—948 (1970).

Cleaver, J. E.: Thymidine metabolism and cell kinetics. In: Neuberger, A., Tatum, E. L. (Eds.): Frontiers of Biology Vol. 6, Amsterdam: North-Holland Publ. Co. 1967.

Darnell, J. E., Pagoulatos, G. N., Lindberg, U., Balint, R.: Studies on the relationship of mRNA to heterogenous nuclear RNA in mammalian cells. Cold. Spr. Harb. Symp. quant. Biol. **35**, 555—560 (1970).

Dick, C., Johns, E. W.: The removal of histones from calf thymus deoxyribonucleoprotein and calf thymus tissue with acetic acid containing fixatives. Biochem. J. **105**, 46 P (1967).

Eckardt, R. A.: Studies on the chromosomal distribution of repetitive nucleotide sequences in polytene chromosomes. J. Cell Biol. **47**, 55a (1970).

French, W. L., Kitzmiller, J. B.: The formation of molecular hybrids *in situ* on chromosomes of *Drosophila melanogaster*. Amer. Zoologist **7**, 782 (1967).

Gall, J. G.: The genes for ribosomal RNA during oögenesis. Genetics **61**, 121—132, Suppl. 1 (1969).

— Mary Lou Pardue: The formation and detection of RNA-DNA hybrid molecules in cytological preparations. Proc. nat. Acad. Sci. (Wash.) **63**, 378—383 (1969).

— — Nucleic acid hybridization in cytological preparations. Methods in Enzymology 12 C. (in press) (1971).

Gillam, I., Millward, S., Blew, D., von Tigerstrom, M., Wimmer, E., Tener, G. M.: The separation of soluble ribonucleic acids on benzoylated Diethylaminoethylcellulose. Biochemistry **6**, 3043—3056 (1967).

Gillespie, D., Spiegelman, S.: A quantitative assay for DNA-RNA hybrids with DNA immobilized on a membrane. J. Mol. Biol. **12**, 829—842 (1965).

Hnilica, L. S., Kappler, H. A., Hnilica, V. S.: Biosynthesis of histones and acidic nuclear proteins under different conditions of growth. Science **150**, 1470—1472 (1965).

John, H. A., Birnstiel, M. L., Jones, K. W.: RNA-DNA hybrids at the cytological level. Nature (Lond.) **223**, 582—587 (1969).

Jones, K. W.: Chromosomal and nuclear location of mouse satellite DNA in individual cells. Nature (Lond.) **225**, 912—915 (1970).

Kaufmann, B. P., Iddles, Marcia K.: Ectopic pairing in Salivary-gland chromosomes of Drosophila. Port. Acta Biol. **7**, 225—248 (1965).

Laird, C. D., McCarthy, B. J.: Magnitude of interspecific nucleotide sequence variability in Drosophila. Genetics **60**, 303—322 (1968a).

— — Nucleotide sequence homology within the genome of *Drosophila melanogaster*. Genetics **60**, 323—324 (1968b).

Lindsley, Dan L., Grell, E. H.: Genetic Variations of *Drosophila melanogaster*. Carnegie Inst. Wash. Publ. No. 627 (1967).

Macgregor, H. C., Kezer, J.: The chromosomal localization of a heavy satellite DNA in the testis of *Plethodon c. cinereus*. Chromosoma **33**, 167—182 (1971).

McCarthy, B. J., Church, R. B.: The specificity of molecular hybridization reactions. Ann. Rev. Biochem. **39**, 131—150 (1970).

Miller, O. L., Beatty, B. R.: Extrachromosomal nucleolar genes in amphibian oöcytes. Genetics **61**, (Suppl. 1) 133—144 (1969).

Pardue, M. L., Gall, J. G.: Chromosomal localization of mouse satellite DNA. Science **168**, 1356—1358 (1970).

— Gerbi, S. A., Eckardt, R. A., Gall, J. G.: Cytological localization of DNA complementary to ribosomal RNA in polytene chromosomes of Diptera. Chromosoma (Berl.) **29**, 268—290 (1970).

Pelling, C.: Ribonukleinsäure-synthese der Riesenchromosomen. Autoradiographische Untersuchungen an *Chironomus tentans*. Chromosoma (Berl.) **15**, 71—122 (1964).

— Puff RNA in polytene chromosomes. Cold Spr. Harb. Symp. quant. Biol. **35**, 521—531 (1970).

Rae, Peter M. M.: Chromosomal distribution of rapidly reannealing DNA in *Drosophila melanogaster*. Proc. nat. Acad. Sci. (Wash.) **67**, 1018—1025 (1970).

Ritossa, F. M., Spiegelman, S.: Localization of DNA complementary to ribosomal RNA in the nucleolus organizer region of *Drosophila melanogaster*. Proc. nat. Acad. Sci. (Wash.) **53**, 737—745 (1965).

— Atwood, K. C., Spiegelman, S.: On the redundancy of DNA complementary to amino acid transfer RNA and its absence from the nucleolar organizer region of *Drosophila melanogaster*. Genetics **54**, 663—676 (1966a).

— — — A molecular explanation of the bobbed mutants of Drosophila as partial deficiencies of "ribosomal" DNA. Genetics **54**, 819—834 (1966b).

Steffensen, Dale M., Wimber, D. E.: Localization of tRNA genes in the salivary chromosomes of Drosophila by RNA:DNA Hybridization. Genetics **69**, 163—178 (1971).

Tartof, K. D., Perry, R. P.: The 5S RNA genes of *Drosophila melanogaster*. J. Mol. Biol. **51**, 171—183 (1970).

Wimber, D. E., Steffensen, D. M.: Localization of 5S RNA genes in Drosophila chromosomes by RNA-DNA hybridization. Science **170**, 639—641 (1970).

Wolstenholme, D. R.: The distribution of RNA and DNA in salivary gland chromosomes of *Chironomus tentans* as revealed by fluorescence microscopy. Chromosoma (Berl.) **17**, 219—229 (1965).

Yasmineh, W. G., Yunis, J. J.: Satellite in DNA mouse heterochromatin. Biochem. Biophys. Res. Commun. **35**, 779—782 (1969).

Nucleic Acid Hybridization and the Nature of Chromosomal Protein Bound RNA

Ru Chih C. Huang and M. Mitchell Smith

Department of Biology, Johns Hopkins University, Baltimore, Maryland

I. Introduction

The past decade has seen a rapid advance in the state of knowledge of isolated interphase chromosomes, termed chromatin, in a variety of higher organisms. A number of relatively standard procedures for the study of such chromosomal material have been developed (Bonner et al., 1968a) and enough of its biological properties are known (Bonner et al., 1968b) that it now appears possible to begin to elucidate the mechanisms involved in gene regulation in higher organisms. In this paper we review recent evidence regarding the structure and function of a regulatory ribonucleoprotein of chromatin.

Two lines of evidence, in one case from the proteins synthesized and in the other from the RNA transcribed, support the concept that genetic template specificity is preserved and may be expressed by isolated chromatin. Bonner, Huang, and Gilden (1963) have coupled chromatin dependent RNA synthesis to a messenger RNA-dependent protein synthesizing system of *E. coli* and found that developing pea cotyledon chromatin directs the synthesis of pea seed reserve globulin. Chromatin isolated from apical buds, which do not synthesize globulin *in vivo*, does not support synthesis of the globulin *in vitro*.

An examination of the chromatin-dependent RNA species synthesized *in vitro* has also been made. Paul and Gilmour (1966, 1968, 1969) and Smith, Church and McCarthy (1969) using RNA-DNA competition hybridization experiments have demonstrated that the hybridizable *in vivo* RNA from a variety of tissues competes extensively with RNA transcribed *in vitro* off the homologous tissue chromatin. The hybridizable chromatin-primed RNA also competes out labelled *in vivo* RNA. These results have been confirmed in our own laboratory using chromatin from 11 day old chick embryos (Huang and Huang, 1969). Because of the hybridization conditions, however, the experiments measure only those RNA species homologous to redundant sequences of the DNA and give no information about species homologous to unique DNA sequences. In addition, because of the imperfect base pairing of the redundant nucleic acids of higher organisms, only the remarkable similarity and not the identity of the natural and *in vitro* chromatin-primed RNAs is established. Nevertheless RNA-DNA hybridization represents the most convenient and usually the only functional assay for chromatin. In contrast to chromatin, purified DNA shows little, if any, specificity of *in vitro* transcription and thus the controlling complex must exist elsewhere in the chromatin material.

It is well known that histones repress the transcription of DNA (Huang and Bonner, 1962; Barr and Butler, 1963; Allfrey, Littau, and Mirsky, 1963; Marushige and Bonner, 1966) and that the removal of histones from isolated chromatin results in increased template activity (Bonner et al., 1968b; Paoletti and Huang, 1969). The histones themselves, however, while demonstrating low level preferences for G-C and A-T regions of DNA (Leng and Felsenfeld, 1966), appear not to have the necessary sequence specificity required for control of transcription (Johns and Butler, 1964) and the types of histones are generally similar in different organisms and in different organs as well (Hnilica, 1968). Evidence of this type prompted the search for an RNA species involved in the regulation of chromatin transcription.

II. Identification and Function of Chromosomal RNA

A unique chromosomal RNA closely associated with chromatin proteins was first reported for pea bud chromatin by Huang and Bonner (1965). If chromatin is isolated from 11 day old chick embryos, for example, which have been labelled with radioactive phosphate, and centrifuged in 4M cesium chloride at high speed for 48 h, the DNA pellets while all the protein forms a skin at the top of the tubes. This skin contains a portion of the radioactivity and may be removed and completely

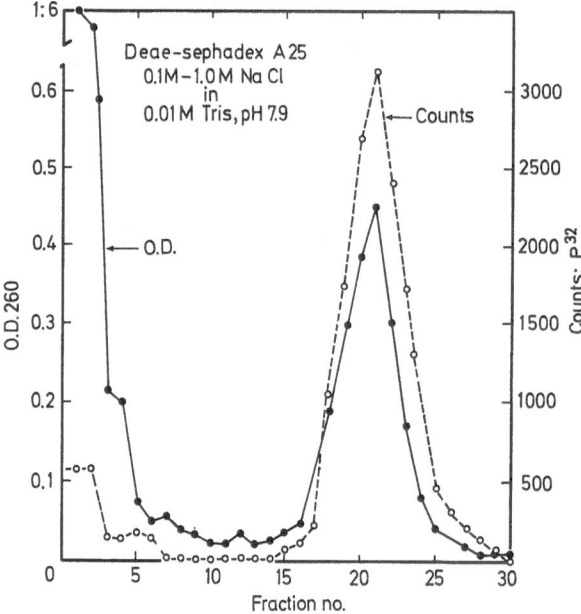

Fig. 1. Purification of pronase digested chromosomal RNA. Eleven day old chick embryos were labelled with ^{32}P-inorganic phosphate by injection into the yolk. Chromatin was prepared (Huang and Huang, 1969) and centrifuged to equibrium in 4M CsCl. The protein skin at the top of the tube was solubilized by extensive pronase digestion and eluted off DEAE-Sephadex A25 with a linear gradient of 0.1M-1.0M NaCl in 0.01M Tris pH 7.9

digested with pronase. When the digested material is eluted off of an A-25 DEAE-sephadex column with a 0.1 M to 1.0 M sodium chloride gradient, one obtains a peak of RNA (Fig. 1) which is completely sensitive to RNAase digestion and is alkali labile. If instead of 4 M cesium chloride, centrifugation is done in 2 M cesium the DNA will still pellet but instead of a protein skin, chromosomal RNA and nuclear proteins may be recovered from a band in the center of the gradient, demonstrating the close association of this RNA with chromatin proteins (HUANG and BONNER, 1965). The RNA-protein complex is not the result of non-specific adsorption during preparation since purified radioactive ribosomal RNA when added to the system pellets with the DNA.

Observations of chromosomal RNA have now been extended to a wide variety of systems including chick embryo (HUANG, 1967), mouse liver and mouse hepatoma (MARZLUFF, SMITH, and HUANG, 1972), rat ascites tumor (DAHMUS and McCONNELL, 1969), and calf thymus (SHIH and BONNER, 1969). All such chromosomal RNAs so far studied are characterized and may be distinguished from other RNA types by their low molecular weight, high content of a dihydropyrimidine, association with chromosomal proteins, and molecular hybridization properties. The base compositions of chromosomal, ribosomal, and transfer RNA from chick embryos are given in Table 1. BONNER and WIDHOLM (1967) have demonstrated by RNA-

Table 1. Base ratio of protein bound RNA from chromatin, ribosomal RNA and transfer RNA from 11 day old chick embryos

Nucleotide bases	Chromatin PB RNA	% Moles Ribosomal RNA	Transfer RNA
Cytidylic acid	24.4	31.9	22.7
Adenylic acid	22.5	24.2	25.8
Guamylic acid	26.4	30.0	27.7
Uridylic acid	16.3	13.8	23.1
Dihydropyrimidine	11.3	< 1	3.1

DNA hybridization experiments using material prepared as just described above, that chromosomal RNA, obtained from pea chromatin hybridizes to about 5% of the genome at saturation; it is confined almost exclusively to the nucleus and, further, it is distinct from either transfer or ribosomal RNA. DAHMUS and McCONNEL (1969) have obtained similar hybridization results using NOVIKOFF ascites cells. In addition they found that the extent of methylation can distinguish chromosomal from transfer RNA and that its rate of synthesis is significantly lower then that of messenger RNA.

Hybridization studies have indicated that the sequence populations of chromosomal RNA change during the development of the chick limb bud (WHITE and HUANG, unpublished results). In the case of peas BONNER and WIDHOLM (1967) found that pea bud chromosomal RNA has some sequences in common with cotyledon RNA but that it also contains some sequences unique to buds. Taken together with its nuclear location such evidence suggests that chromosomal RNA may have some regulatory function. A more critical approach to determining the functional properties of chromosomal RNA, however, has been to examine the template specificity, through RNA-DNA hybridization, of chromatin reconstituted in the presence or absence of this protein-bound RNA component.

When chromatin from 11 day old chick embryos is dissolved in 2M guanidine-HCl, all of the chromosomal RNA and histones and most of the non-histone chromatin proteins are completely dissociated. The DNA may be pelleted by high speed centrifugation if desired. If instead, the ionic strength is reduced gradually by stepwise dialysis against 1.0, 0.8, 0.6, 0.5, 0.4, 0.3 and 0.1 M GuCl for 4 h each in the continued presence of 5M urea and the urea is then dialyzed out, the dissociated chromatin will be reconstituted. The thermal denaturation profile of the reconstituted chromatin is very close to that of native chromatin giving T_m values within 2° of each other. When RNA is transcribed off such a reconstituted chromatin *in vitro*, using *E. coli* RNA polymerase, it may be completely competed out by hybridization with increasing amounts of unlabelled homologous nuclear RNA made *in vivo* (Fig. 2). Reconstituted chromatin, then, appears to preserve its natural transcription specificity just as native chromatin does.

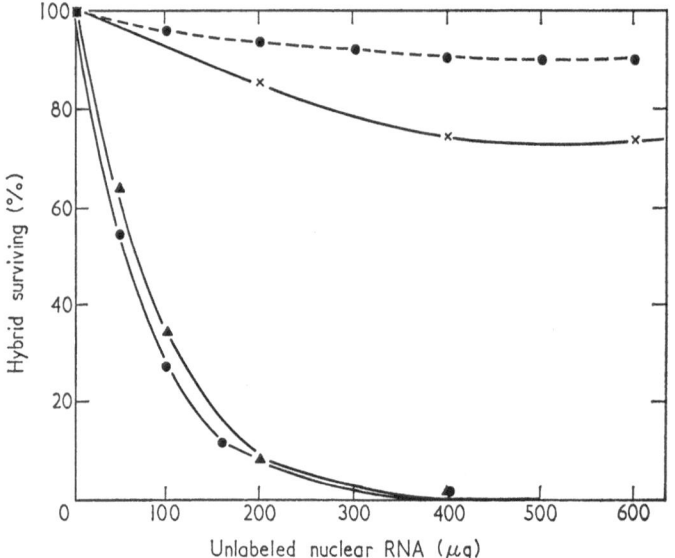

Fig. 2. Hybridization competition between unlabeled nuclear RNA and RNA primed *in vitro* by native chromatin, reconstituted chromatin with or without $Zn(NO_3)_2$ treatment. Nuclear RNA competes with RNA primed by native chromatin and reconstituted up to 100% or so, but only to approximately 26% with RNA primed by $Zn(NO_3)_2$-treated, reconstituted chromatin. Reconstitution, in the absence of urea, results in chromatin which yields non-specific RNA which competes poorly with nuclear RNA. In these experiments, 50 µg of DNA were used for each of 25 µg of ^{32}P-labeled RNA synthesized *in vitro*. Hybridization was carried out in 4 × SSC at 67° C for 18 h. —— ● —— ●, RNA primed by chromatin reconstituted without urea; — × — ×, RNA primed by $Zn(NO_3)_2$-treated chromatin, reconstituted with urea; — ▲ — ▲, RNA primed by chromatin reconstituted with urea; ——●——, RNA primed by native chromatin (From Huang and Huang, 1969)

If urea is omitted during the dialysis to low salt, the resulting reconstituted chromatin does not show a normal melting profile and in addition the RNA transcribed is only poorly competed by unlabelled nuclear RNA (Fig. 2). Thus, reconstitution in the absence of urea results in chromatin which does not have specific transcription and produces RNA copies not produced by the tissue of origin. The reasons for this are not yet clear. Perhaps urea permits a relaxation of the DNA duplex allowing a more correct pairing interaction with the RNA or perhaps it permits correct protein-protein interactions during reassociation.

If the protein-bound chromosomal RNA does indeed control the specificity of gene transcription and it was hydrolyzed away prior to reconstitution of the chromatin, one might anticipate that such chromatin would be non-specific. The RNA transcribed would represent random regions of the genome and thus would show poor competition with homologous *in vivo* RNA. On the other hand, RNA transcribed off the whole genome of purified DNA should compete more efficiently with the non-specific chromatin-primed RNA than with native chromatin primed RNA.

BUTZOW and EICHORN (1965) have demonstrated that zinc nitrate is capable of hydrolyzing RNA but not DNA in a variety of salt conditions. Chick chromatin was thus dissociated in 2 M GuCl and 5 M urea and treated with 10^{-4} M zinc nitrate at 37° for 20 h. This treatment can be shown to hydrolyze chromosomal RNA while not affecting the DNA or the nuclear proteins as judged by polyacrylamide electrophoresis patterns. The zinc salt was dialyzed away and the chromatin reconstituted as usual in the presence of urea. Under such conditions what in fact is discovered is that now the reconstituted chromatin no longer retains its template specificity. The RNA transcribed *in vitro* shows almost no reduction of hybrid formation with added nuclear RNA made *in vivo* (Fig. 2). On the other hand, RNA transcribed off the entire genome of purified DNA competes much more efficiently against zinc-treated chromatin-primed RNA than against native chromatin-primed RNA (Fig. 3).

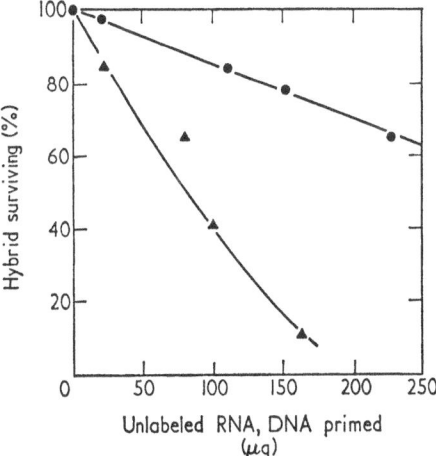

Fig. 3. Competition between RNA primed by native DNA and reconstituted chromatin-primed RNA in forming hybrids with DNA. Hybridization conditions were the same as in Fig. 2. —●—●—, RNA primed by reconstituted chromatin; —▲—▲—, RNA primed by zinc-treated, reconstituted chromatin (From HUANG and HUANG, 1969)

Identical results have been obtained by Bekhor, Kung, and Bonner (1969) for peas and using RNAase instead of zinc nitrate to hydrolyze the RNA. Together such evidence strongly supports the hypothesis that protein-bound chromosomal RNA is a main factor for the regulation of the pattern of genetic transcription.

III. Structure of Chromosomal Protein Bound RNA (PBRNA)

Although most of the studies of isolated chromosomal RNA have used the RNA digested free of protein by pronase, it is the protein bound RNA complex that is the functional unit in chromatin. The determination of the structure of this material, however, has proved to be a most difficult problem. Important aspects of the structure of chick embryo chromosomal RNA have only recently been solved (Marzluff, Smith, and Huang, 1972).

As mentioned in the previous section, if chick embryo chromatin is dissolved in 2.5 M guanidine-HCl buffered at pH 7.8 with 0.01 M Tris, the DNA may be pelleted by high speed centrifugation leaving the chromatin proteins and PBRNA in the supernatant. The PBRNA, because of its greater density, may be separated from the other components by adjusting the supernatant to 4M cesium chloride, 20%

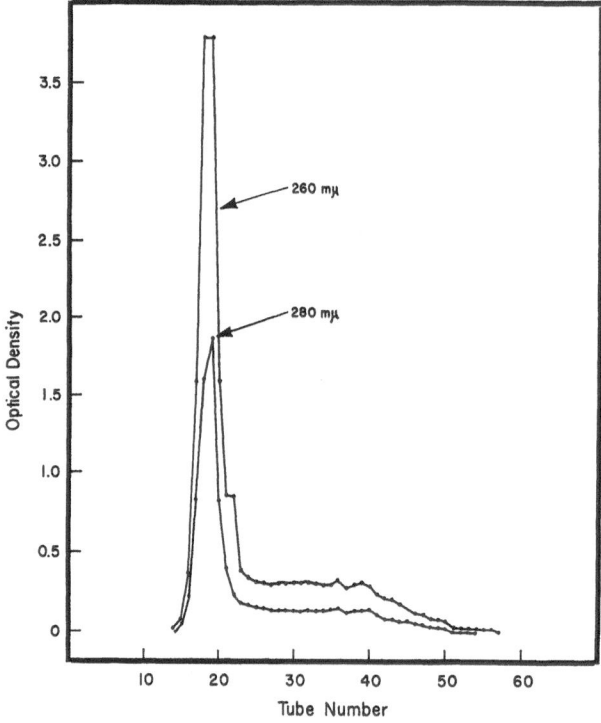

Fig. 4. Preparative Disc Electrophoresis of PBRNA. Chromatin was dissolved in 2.5 M guanidine hydrochloride buffered at pH 7.8 with 0.01 M Tris and the DNA pelleted by centrifugation at $105\,000 \times g$ for 24 h. The supernatant was adjusted to 4 M CsCl, 2.0 M GuCl and centrifuged to equilibrium. The RNA rich lower portion of the tubes were collected and dialyzed to 0.001 M Tris pH 7.9. The PBRNA was then electrophoresed from cathode to anode, on a Canalco Preparative Disc Electrophoresis apparatus

guanidine-HCl, and centrifuging to equilibrium. When the tubes are pierced and fractions collected, 85—90% of the PBRNA is recovered in the bottom ten fractions of the tubes. This material may be combined and further purified by electrophoresis in polyacrylamide gel on a Canalco Preparative Disc Electrophoresis unit. The sample moves from cathode to anode as a rapidly migrating homogeneous fraction with a protein/RNA mass ratio of 0.39 and a molecular weight of 17000 (Fig. 4). It is this purified PBRNA that was used for structural analysis.

The determination of the mechanism of linkage between the RNA and protein portions of PBRNA was initiated by isolating the fragments remaining after digestion of the RNA. To preserve the structure of the dihydropyrimidine, alkaline hydrolysis was avoided (JANION and SHUGAR, 1960), and the PBRNA sample was digested by a mixture of pancreatic, T1 and T2 ribonucleases. The fragments were separated according to their sizes. Chromatography of the digest on a Biogel P6 column separates a fraction containing polypeptide material bound to a dihydropyrimidine (polypeptide-H_2Py), with an approximate molecular weight of 3000, from a second peak containing a fraction of small peptides, two or three amino acids long, linked to a dihydropyrimidine (peptide-H_2Py), a fraction of a single amino acid linked to a dihydropyrimidine (amino acid-H_2Py), and free nucleotides. The peptide-H_2Py and amino acid-H_2Py were separated from free nucleotides and each other on a Biogel P2 column followed by ion exchange chromatography on Dowex 1 × 8.

Base ratio determinations show PBRNA to contain about 11.3% dihydropyrimidine. Of this, about 35% was bound in the polypeptide-H_2Py fraction, 22 % in the peptide-H_2Py fraction, and 42% in the amino acid-H_2Py fraction. Very little free dihydropyrimidine was detected and all protein-RNA complexes involved the dihydropyrimidine base. Amino acid analyses on all three fractions indicate that the polypeptide is not basic in nature and that it is likely that the only amino acid bound to the dihydropyrimidine in amino acid-H_2Py is serine (Table 2).

Table 2. Amino acid composition of polypeptide-H_2Py, peptide-H_2Py, and amino acid-H_2Py (mole %)

	Polypeptide-H_2Py	Peptide-H_2Py	Amino acid-H_2Py
lysine	7.4	9.1	0
histidine	1.1	0	0
arginine	1.8	0	0
aspartic	8.0	1.9	0
glutamic	18.0	8.4	0
threonine	3.3	0	0
serine	18.3	12.0	32.1
glycine	23.0	7.7	13.2
proline	2.4	0	0
alanine	7.4	3.8	0
cystine	0.5	0	0
methionine	0.2	0	0
valine	2.4	0	0
isoleucine	1.6	11.6	0
leucine	1.8	9.6	0
tyrosine	1.1	0	0
phenylalanine	1.8	0	0
H_2Py	3.3	36.1	54.7

The detailed structural linkage studies were done on the amino acid-H_2Py fraction. Hydrolysis of this material in 6 N HCl at 110° C for 18 h gives the degradation products ammonia, serine, and β-amino isobutyric acid. The β-amino isobutyric acid was identified by comparison with the authentic compound upon thin layer chromatography in four different solvent systems and gas liquid chromatography.

This information suggests that the structure of amino acid-H_2Py might be serine, 5-methyl, 5,6-dihydrocytosine (Fig. 5). In this structure, the exocyclic nitrogen of 5-methyl dihydrocytosine is bound to the α-carboxyl group of serine forming an amide bond. The exact linkage in the other two fractions is not precisely known, although it is assumed to be the same. If, for example, peptide-H_2Py or polypeptide-H_2Py are hydrolyzed under mild conditions where the structure of amino acid-H_2Py is known to be stable but in which peptide bonds are broken, their ultraviolet absorption spectra gradually approach that of amino acid-H_2Py. Thus, chromosomal PBRNA is pictured as a continuous RNA molecule of 40 to 50 nucleotides containing several 5-methyl, dihydrocytosine bases which are bound to either the single amino acid serine, or peptide of two to three amino acids, or a polypeptide of 30 to 40 amino acids (Fig. 5).

Shih and Bonner (1969) have reported β-alanine as a decomposition product of chromosomal RNA from calf thymus chromatin and Jacobson and Bonner (1969) have found β-amino isobutyric acid as a decomposition product in the case of rat ascites tumor PBRNA. These authors have suggested that the parent dihydro-

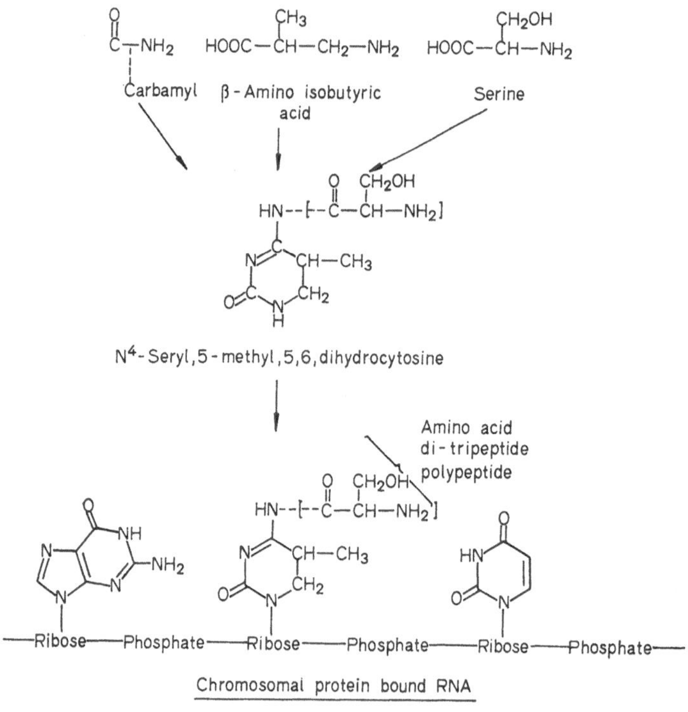

Fig. 5. Postulated structures of chromosomal protein bound RNA. They suggest that the parent dihydropyrimidine may be bound to amino acid.

pyrimidine compounds are dihydrouridylic acid and dihydrothymidylic acid respectively. Clearly, dihydrocytidylic acid and 5-methyl, dihydrocytidylic acid will give these same decomposition products.

IV. Hybridization Properties of Chromosomal PBRNA

Very little is known concerning the mechanism of action of chromosomal PBRNA. Presumably it must be capable of recognizing DNA in a sequence-specific manner and also influence the assembly of histones and possibly other nuclear proteins. BEKHOR, BONNER, and DAHMUS (1969) have reported the binding of chromosomal RNA to native double stranded DNA. The RNA was isolated by pronase digestion and phenol extraction of the protein skin formed upon centrifugation of Novikoff ascites tumor chromatin in 4M CsCl, followed by elution off DEAE Sephadex A-25 as described earlier. The resulting material is the bare RNA free of most of the bound protein.

This RNA is reported to hybridize with native double stranded DNA during incubation for 24 h at $1°$ C in the presence of $1 \times$ SSC and 4M urea. It should be noted that these conditions provide for the reaction to occur better than $50°$ C below the T_m of the native DNA which may not be conducive to highest specificity (McCo-NAUGHY, LAIRD, and McCARTHY, 1969).

The hybrids formed are stable in sucrose density gradient sedimentation, cesium chloride equilibrium centrifugation, and are retained by hydroxylapatite up to $80—90°$ C at which point they elute with the melting DNA. The reaction is reportedly specific for homologous rat DNA, and although some hybridization occurs with calf DNA these hybrids have thermal stabilities approximately $20°$ C lower. The percentage hybridization values reported for the homologous reaction vary from 1.0 to about 5.6% but it is not clear whether identical input ratios were used in each case. "Messenger RNA", that is, RNA transcribed off rat ascites chromatin *in vitro* by *E. coli* RNA polymerase, does not hybridize at all with native DNA under the same conditions. Of potential significance is the finding that the T_m of hybrids formed between chromosomal RNA and native double stranded DNA is $12°$ C higher than those formed with single stranded denatured DNA under identical reaction conditions. It may be that this is not a valid comparison, however, since the conditions for obtaining optimum thermal stability may be different for the two cases. No data are available on this question. Finally, although chromosomal RNA may hybridize to native DNA it is clear that the situation *in vivo*, when the RNA is covalently linked to protein, is more complex. For example, the RNA used for the hybridizations was originally separated from the chromatin DNA by extraction with 4M CsCl and yet hybrids formed were stable at 5.68 molar. So far, all attempts in this laboratory to hybridize native undigested chick embryo chromosomal PBRNA to purified native double stranded DNA have been unsuccessful.

A property of the RNA-protein linkage suggested here is that it would not disrupt hybridization of the molecule to single stranded DNA. Fig. 6 shows a picture of N-seryl-5-methyl-5,6-dihydrocytosine paired with guanine built with space filling CPK atomic models and demonstrates that the pair is sterically permissible. Open ring forms of dihydrouracil or dihydroribothymine when linked to an amino acid through the 4-carbon would disrupt normal base pair hydrogen bonds.

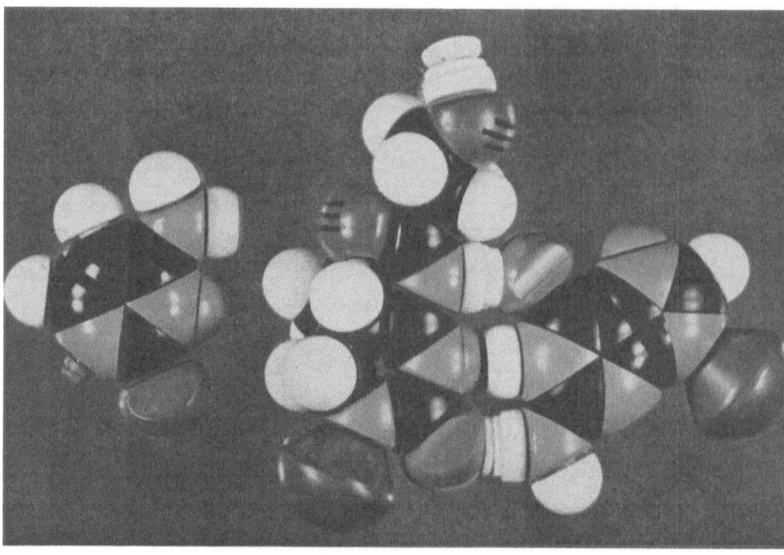

Fig. 6. A guanine = N⁴-seryl-5-methyl-5,6-dihydrocytosine base pair (right) constructed from CPK atomic models. At left is shown a cytosine molecule for comparison

It has recently been suggested (Mayfield and Bonner, 1972) that chromosomal PBRNA is the hypothetical activator RNA of the gene control model of Britten and Davidson (1969). Although attractive, there is as yet no direct evidence to support the suggestion that the molecule functions in this manner. Certainly PBRNA is homologous to repetitious sequences of the genome since hybridization is detectable with the usual filter techniques. It is not known, however, whether or not PBRNA is exclusively transcribed by repetitious sequences.

Sivolap and Bonner (1971) separated pea DNA on hydroxylapatite into repetitive middle repetitive, and unique sequence fractions on the basis of reassociation kinetics. Pea seed bud chromosomal PBRNA was found to hybridize extensively to repetitive DNA filters, much less to middle repetitive DNA, and negligibly to unique sequence DNA. However, with specific activities of chromosomal PBRNA of only 330 cpm/μg, there could be PBRNA sequences homologous with over a hundred thousand different unique DNA sequences and even assuming the reactions could be driven to completion less than 10 cpm of hybrid would be detected on a 10 μg DNA filter. Clearly the homology of PBRNA to unique DNA sequences remains to be determined.

Many structural and functional studies on this class of chromosomal ribonucleoprotein molecules remain to be done. The exact RNA-protein linkage needs to be confirmed by chemical synthesis from known compounds. The biosynthesis and metabolism of the molecules is largely unstudied at this time. However, the resolution of these and other questions promises to provide significant insights into the mechanism of genetic control in eukaryotes.

This work was supported by grants from the National Institute of Health, Grant GM 13733, American Cancer Society ACS Md 69-70.

References

ALLFREY, V. G., LITTAU, V. C., MIRSKY, A. E.: On the role of histones in regulating ribonucleic acid synthesis in the cell nucleus. Proc. nat. Acad. Sci. (Wash.) **49**, 414 (1963).

BARR, G. C., BUTLER, J. A. V.: Histones and gene function. Nature (Lond.) **199**, 1170 (1963).

BEKHOR, I., BONNER, A., DAHMUS, G.: Proc. nat. Acad. Sci. (Wash. **62**, 27 (1969).

— KUNG, G., BONNER, J.: Sequence-specific interaction of DNA and chromosomal protein. J. molec. Biol. **39**, 351 (1969).

BONNER, J., HUANG, R. C., GILDEN, R.: Chromosomally directed protein synthesis. Proc. nat. Acad. Sci. (Wash.) **50**, 893 (1963).

— WIDHOLM, J.: Molecular complementarity between nuclear DNA and organ-specific chromosomal RNA. Proc. nat. Acad. Sci. (Wash.) **57**, 1379 (1967).

— CHALKLEY, G. R., DAHMUS, M., FAMBROUGH, D., FUJIMURA, F., HUANG, R. C., HUBERMAN, J., JENSEN, R., MARUSHIGE, K., OHLENBUSCH, H., OLIVERA, B., WIDHOLM, J.: Isolation and characterization of chromosomal nucleoproteins. In: GROSSMAN, L., MOLDAVE, K. (Eds.): Methods in Enzymology (Nucleic Acids), Vol. XII, part B, p. 3—64. New York: Academic Press 1968a.

— DAHMUS, M. E., FAMBROUGH, D., HUANG, R. C., MARUSHIG, K., TUAN, D. Y. H.: The biology of isolated chromatin. Science **159**, 47—56 (1968b).

BRITTEN, R. J., DAVIDSON, E. H.: Gene Regulation for Higher Cells: A Theory. Science **165**, 349 (1969).

BUTZOW, J., EICHHORN, G.: Interactions of metal ions with polynucleotides and related compounds. IV. Degradation of polyribonucleotides by zinc and other divalent metal ions. Biopolymers **3**, 95 (1965).

DAHMUS, M., McCONNELL, D.: Chromosomal ribonucleic acid of rat ascites cells. Biochemistry **8**, 1524—1534 (1969).

HNILICA, L. S.: Proteins of the cell nucleus. In: DAVIDSON, J. N., COHN, W. E. (Eds.): Progress in Nucleic Acid Research and Molecular Biology, Vol. 7. New York: Academic Press 1968.

HUANG, R. C., BONNER, J.: Histone, a suppressor of chromosomal RNA synthesis. Proc. nat. Acad. Sci. (Wash.) **48**, 1216 (1962).

— — Histone-bound RNA, a component of native nucleohistone. Proc. nat. Acad. Sci. (Wash.) **54**, 960—967 (1965).

— Dihydrouridylic acid containing RNA from chick embryo chromatin. Fed. Proc. **26**, 603 1933 A (1967).

— Control of transcription in higher organism. In: LI, H. W., TS'O, P. O., HUANG, P. C., KUO, T. T. (Eds.): Symposium on Recent Development in Biochemistry, p. 133. Taipei, Taiwan: Academic Sinica 1968.

— HUANG, P. C.: Effect of protein-bound RNA associated with chick embryo chromatin on template specificity of the chromatin. J. molec. Biol. **39**, 365—378 (1969).

JACOBSON, R. A., BONNER, J.: The occurrance of dihydro-ribothymidine in chromosomal RNA. Biochem. Biophys. Res. Commun. **33**, 716 (1968).

JANION, C., SHUGAR, D.: Absorption spectra, structure, and behavior towards some enzymes of dihydropyrimidines and dihydro-oligonucleotides. Acta Biochim. Polon. **7**, 309 (1960).

JOHNS, E. W., BUTLER, J. A. V.: Specificity of the interactions between histones and deoxyribonucleic acid. Nature (Lond.) **204**, 853 (1964).

LENG, M., FELSENFELD, G.: The preferential interactions of polylysine and polyarginine with specific base sequences in DNA. Proc. nat. Acad. Sci. (Wash.) **56**, 1325 (1966).

MARMUR, J., DOTY, P.: Heterogeneity of deoxyribonucleic acids. I. dependence on composition of the configurational stability of deoxyribonucleic acids. Nature (Lond.) **183**, 1427 (1959).

MARUSHIGE, K., BONNER, J.: Template properties of liver chromatin. J. molec. Biol. **15**, 160 (1966).

MARZLUFF, W., SMITH, M. M., HUANG, R. C.: Characterization of peptide-bound RNA from chick embryo chromatin. Fed. Proc. (abstract) (1972).

Mayfield, J. E., Bonner, J.: A partial sequence of nuclear events in regenerating rat liver. Proc. Nat. Acad. Sci. (Wash.) **69**, 7 (1972).

Paoletti, R. A., Huang, R. C.: Characterization of sea urchin sperm chromatin and its basic proteins. Biochemistry **8**, 1615—1625 (1969).

Paul, J., Gilmour, R.: Template activity of DNA is restricted in chromatin. J. molec. Biol. **16**, 242 (1966).

— — Organ-specific restriction of transcription in mammalian chromatin. J. molec. Biol. **34**, 305 (1968).

— — RNA transcribed from reconstituted nucleoprotein is similar to natural RNA. J. molec. Biol. **40**, 137 (1969).

Shih, T., Bonner, J.: Chromosomal RNA of calf thymus chromatin. Biochem. biophys. Acta (Amst.) **182**, 30 (1969).

Sivolap, Y. M., Bonner, J.: Association of chromosomal RNA with repetitive DNA. Proc. Nat. Acad. Sci. (Wash.) **68**, 387 (1971).

Smith, K., Church, R., McCarthy, B.: Template specificity of isolated chromatin. Biochemistry **8**, 4271 (1969).